Rudolph

Intuitive Analog Electronics

AF769863

McGraw-Hill Series in Intuitive IC Electronics

This series will help the reader gain an intuitive understanding of electronics and computers. Mathematics is kept to a minimum, as the reader gets "inside" the devices and circuits to grasp, from the electron level up, the workings of integrated circuits, digital computers, operational amplifiers, and other electronics-related topics. The following volumes are planned for the series, and each one is written by Thomas M. Frederiksen, whose Intuitive IC Electronics (McGraw-Hill) has proved popular with engineers, managers, students, and hobbyists worldwide.

INTUITIVE DIGITAL COMPUTER BASICS: An Introduction to the Digital World (1988)

INTUITIVE ANALOG ELECTRONICS: From Electron to Op Amp (1988)

INTUITIVE OPERATIONAL AMPLIFIERS: From Basics to Useful Applications (1988)

INTUITIVE CMOS ELECTRONICS: The Revolution in VLSI, Processing, Packaging, and Design (1989)

INTUITIVE IC ELECTRONICS, Second Edition (1989)

For more information about other McGraw-Hill materials, call 1-800-2-MCGRAW in the United States. In other countries, call your nearest McGraw-Hill office.

Intuitive Analog Electronics

From Electron to Op Amp

Thomas M. Frederiksen

McGraw-Hill Book Company

New York St. Louis San Francisco Auckland
Bogotá Hamburg London Madrid Mexico
Milan Montreal New Delhi Panama
Paris São Paulo Singapore
Sydney Tokyo Toronto

Library of Congress Cataloging-in-Publication Data

Frederiksen, Thomas M.
Intuitive analog electronics from electron to op amp.

Includes index.
1. Analog electronic systems. I. Title.
TK7867.F74 1988 621.381 87-35358
ISBN 0-07-021962-1
ISBN 0-07-021963-X (pbk.)

Copyright © 1989 by McGraw-Hill, Inc. All rights reserved. Printed in the United States of America. Except as permitted under the United States Copyright Act of 1976, no part of this publication may be reproduced or distributed in any form or by any means, or stored in a data base or retrieval system, without the prior written permission of the publisher.

1234567890 DOC/DOC 8954321098

ISBN 0-07-021962-1

ISBN 0-07-021963-X {PBK.}

The editors for this book were Daniel A. Gonneau and Nancy Young, the designer was Naomi Auerbach, and the production supervisor was Richard A. Ausburn. It was set in Century Schoolbook. It was composed by the McGraw-Hill Book Company Professional & Reference Division Composition Unit.

Printed and Bound by R. R. Donnelley & Sons Company

For more information about other McGraw-Hill materials, call 1-800-2-MCGRAW in the United States. In other countries, call your nearest McGraw-Hill office.

Information contained in this work has been obtained by McGraw-Hill, Inc. from sources believed to be reliable. However, neither McGraw-Hill nor its authors guarantees the accuracy or completeness of any information published herein and neither McGraw-Hill nor its authors shall be responsible for any errors, omissions or damages arising out of use of this information. This work is published with the understanding that McGraw-Hill and its authors are supplying information but are not attempting to render engineering or other professional services. If such services are required, the assistance of an appropriate professional should be sought.

This book is dedicated to the many people, with various occupational specialties, who find themselves thrust into this modern electronic age and who would like to end "engineering intimidation" and simply find out what it's all about.

Contents

Preface

The purpose of this book is to provide an understanding of the concepts used in electronics, especially analog or linear electronics, so that nonelectronic people can appreciate the technical side of their jobs and also all the electronic devices that are appearing in modern life. We are in the golden years of electronics and, by acquiring some background material, any interested person can more fully appreciate the exciting wild growth of electronics.

We are immersed in electronics. Power failures today shut down homes and even entire cities. Nothing works during a power blackout because our society continues to rely more and more on electronics for just about everything. We carry electronics with us on our wrists, in our pockets, and for the earphones that are becoming fashionable headgear for the youth, and some of us even carry electronic packages inside our bodies—where they trigger our hearts.

Electronics gets us up in the morning, fixes our breakfast, and controls our automobiles (and all other forms of transportation except walking and bicycling). We go to our electronic offices; use electronic mail; are entertained by electronic games, TV, and radio; and are kept informed by electronic wire services. We balance our bank accounts, run our factories and businesses (and some of us our homes) with electronic help. Even our national defense, and offense, is electronic these days. Yet, *What is all this electronics and why does it remain so mysterious to so many people?*

A major division of electronics usually separates the *analog* (or linear) systems from the *digital* (the on-off, or binary) computer systems. We have also adopted this separation and this book introduces the concepts that are basic to analog systems. A companion book, *Intuitive Digital Computer Basics,* is also available that similarly explains the basics of electronics and then leads the reader in the direction of digital concepts and digital computers. To assist the learning processes, a commercially available electronic projects kit and a digital multimeter have been integrated into the text to supply a hands-on "lab experience" as a part of each chapter. The only prerequisite for starting with this book is a curious mind. The rewards of comprehension and the enthusiasm that is generated by learning new material will provide the motivation to finish the book.

Chapter 1 starts with some general comments on matter and the very important concept of the electron. The intent is to show the many useful things that are based on electrons. A short look at chemistry then provides the background that is needed to appreciate semiconductors, the basis of solid-state devices. Both the scientific and the engineering representations for large and small numbers are then introduced to supply some background for the few calculations that will come up later. This chapter ends with an introduction to the electronic projects kit, which then becomes the lab kit for the reader and is used in each of the following chapters.

The notions of voltage, cells, and batteries are covered in Chapter 2. The digital multimeter is then used to measure the dc voltages that are available in the lab kit. A number of the different ways that electricity can be generated are also presented.

Chapter 3 introduces the concept of the voltage analogs that are used in analog systems. Some examples of common linear systems are given and electronic components are lightly introduced, complete with the symbols that are used to represent these components. Transducers are also discussed. A definition of linearity is given to help understand the meaning of a linear system or a linear amplifier.

Circuit components operating with direct current are the topic of Chapter 4, the largest chapter. A novel carbon film resistor and a plastic or paper dielectric capacitor, which the reader can easily make out of commonly available materials, are described and the resulting component values are measured. This provides some insight into the problems of manufacturing components and gives an appreciation for large-valued capacitors.

PN semiconductor diodes are described and measurements are made of both forward and reverse characteristics. The base-emitter junctions of the bipolar transistors that are available in the lab kit are used to demonstrate avalanche (zener) diodes. A way is shown to use a standard hair dryer to provide a first-hand indication of the effects that temperature has on diodes, transistors, and resistors.

A large section of this chapter is devoted to bipolar transistors. "Transistor action" is demonstrated with a couple of experiments that make use of the lab kit. Measurements of the current gain, β, of a transistor are also made. The functioning of transistors in amplifiers and also in simple switching circuits are demonstrated using experimental circuits. This chapter ends with a discussion and some experiments on a silicon-controlled rectifier (SCR) that is constructed out of an interconnection of an NPN and a PNP transistor. The importance of the SCR as an undesired parasitic device within both linear and digital integrated circuits is also discussed.

AC circuits are the subject of Chapter 5. Sine waves and oscilloscopes are discussed. AC amplifiers are demonstrated with experiments that make use of the lab kit. A useful ac test oscillator is also constructed. This test oscillator is then used to support many interesting experiments.

In Chapter 6, components are reintroduced; this time they operate with the ac voltage that is supplied by the previously constructed test oscillator. A technique to measure the value of the homemade capacitor is given. Inductors, resonance, and transformers are described and many experimental circuits are used for lab demonstrations.

The op amp, the versatile linear building block, is the subject of Chapter 7. The basic idea of feedback is introduced as "going up an attenuator." The internal functioning of an op amp is described, and the four transistors of the lab kit are used to construct a basic, but useful, demonstration op amp. The input currents, offset voltage, and voltage gain of this simplified op amp are then measured using the digital multimeter. This provides familiarity with the meaning of op amp data sheet specs. Many op amp application circuits are then constructed, such as a voltage follower, both a noninverting and an inverting gain amplifier, a multi-input summing amplifier, a voltage sweep circuit, and a square wave oscillator. The function of an op amp as part of a series voltage regulator is then discussed and the differences between an op amp and a voltage comparator are described.

The final chapter, Chapter 8, discusses the future of analog electronics. Analog, digital, and sampled-data systems are defined and contrasted. The new digital signal processing (DSP) systems are also introduced. The future role of analog electronics, as part of a complex semicustom digital VLSI chip, is mentioned.

Although analog systems are not currently as much in the limelight as digital systems (microprocessors and computers), the continual rapid increase and the pervasiveness of electronics is guaranteeing many applications for analog systems and circuits for many years to come.

I would like to acknowledge the support that I have received from the many nonelectronic people who have attended my seminars and who encouraged me to write this book.

Thomas M. Frederiksen

Chapter

1

Some Preliminary Topics

This chapter provides a basic background in many areas that are necessary to aid the understanding of analog electronics. The word "electronics" is built around the magic "electron." So we will start by taking a look at the electron.

Much is missed in the study of electronics without the ability to do some simple mathematical calculations. The new thing in electronic circuits is the wide range of the numbers that are involved in the calculations. These numbers range from extremely small numbers (such as a millionth part of a millionth part of 1) to very large numbers (such as 10 million), so a way of working with this wide range of numbers in a manageable form will be presented. Finally, an electronic-project lab kit, which is a major support for this book, will be introduced. This kit provides the hands-on familiarity that is necessary to make many of the concepts in electronics seem real to a student.

What Makes Up the World?

Today everybody is hearing about electronics. Electronics implies things that make use of electrons. The natural question is, "What are electrons, and why are they so useful in electronics?" Let's start by considering electrons and how they are involved in the composition of the objects that comprise our physical world.

In the old days, many of the philosophers thought that the world was composed of fire, air, water, and earth, and all things in the physical world were thought to be made out of these basic "elements." As science advanced, it was found that this was not really true, but yet, the significant idea was there that physical things are built up from a rather small number of elemental things. These elements cannot be broken down into other substances by chemical means.

Only 92 elements occur naturally on earth. Today, chemists work with these elements, and they are the basic materials that make up all of the chemical compounds. At present, no elements are found elsewhere in the universe that cannot be accounted for here on earth. There are many different types of elements that range from the common metals, such as iron, copper, zinc, and aluminum, to elements that are gases in their natural state, such as oxygen.

It is the electrons that are associated with the atoms that are used in semiconductor electronics. To feel good about electronics, and especially semiconductor electronics, it is necessary to take a little time to look at these electrons and to find out how they are used in electronic circuits.

The electron has also been useful when it exists in vacuums, for example, in a vacuum tube, which is an amplifying device contained within a glass bottle from which the air inside has been removed. These glass-evacuated bottles were used in electronics in the vacuum-tube days.

Television sets still make use of a strangely shaped bottle, the front side of which forms the viewing screen for the television set. Electrons in the TV tube of a black and white set are directed against the backside of this viewing screen and provide the energy that causes a phosphor coating on the viewing screen to light up. As a very small "spotlight" beam of these electrons is repetitively scanned across the backside of these phosphors, the number of electrons contained within this beam is varied or cut off to provide various light intensities that range from a completely dark screen to the most intense white. This sequence of light and dark areas forms the image.

Color TV tubes provide three different localized small-dot clusters of colored phosphors on the backside of a viewing screen. When these dots are selectively excited by one of the three electron beams, they create the primary colors: red, blue, and green. The relative intensities of these very small primary color dots blend together to create the wide range of colors that are seen on a color TV. The picture tube in television sets is one of the last high-volume applications for vacuum-tube electronics. Most of the rest of the electronics in a TV set have been switched over to the more-efficient, and smaller-sized, solid-state electronics.

Things more elemental than the elements

Each element contains a specific number of electrons; carbon has 12, oxygen has 16, and hydrogen has only 1. Electrons are interesting because they have a relatively small mass (so the mass or the weight of an object is not very largely affected by the number of electrons that it contains), but the important thing is that the number of electrons that are associated with each element controls the formation of chemical compounds.

As the temperature changes, chemical compounds (specific groupings of elements) will change the physical state in which they exist. A common example is water. Water is a chemical compound that exists in the liquid state and is made up of the gaseous elements hydrogen and oxygen. A strange thing to observe is that, when these two gases chemically combine (this very rapid combining creates an explosion), they form a liquid: water. Now, you've seen water exist in a solid form when the temperature is very low, below 32°F; this is ice. At room temperature water is a liquid, and yet, at higher temperatures, water turns into steam, the gaseous phase of water. When the temperature of a compound is raised, this added thermal energy can eventually overcome the electrical attractive forces that hold a compound rigidly together. Thus, solids turn to liquids as the temperature is increased.

Water is a chemical compound that is composed of elements, and each element in turn is composed of protons, neutrons, and electrons. The protons and neutrons contain most of the mass of the elements but occupy only a relatively small total volume. The small spot in which the mass of an element is concentrated, the "nucleus" of the atom, is comprised of these protons and neutrons. Surrounding this nucleus, at relatively large spacings, are layers, or "shells," of electrons. These electrons exist in "cloudlike" formations surrounding the nucleus. The outer layer of electrons affects the way elements react with each other.

Figure 1.1 shows a model of one of the simplest elements: the hydrogen atom. At the center of this atom is the nucleus which has one proton with a positive charge of +1. The nucleus of all atoms is where the mass, the fundamental thing that provides the weight of the atom (and all material things), is concentrated. It is interesting that the diameter of this relatively heavy nucleus is very, very small, and it is also an extremely small percentage of the total atomic diameter. The atomic diameter is established by the shells or layers of the electrons that are in orbit surrounding the central nucleus. This is very similar to our solar system in which the planets are in orbit circling about the center of this system, our sun. In the hydrogen atom, there is only a single electron in orbit around the nucleus.

If electrons are removed from an atom, an unbalance exists in the natural, balanced, neutral state of the charges. In the neutral state, all of the positive charge of the protons within the nucleus is balanced by the total negative charge of the surrounding orbiting electrons. But, if some of these electrons are removed, the atom will end up with a positive charge in this unbalanced state, and this atom then becomes a positively charged "ion." An ion has an electrical charge (it can be

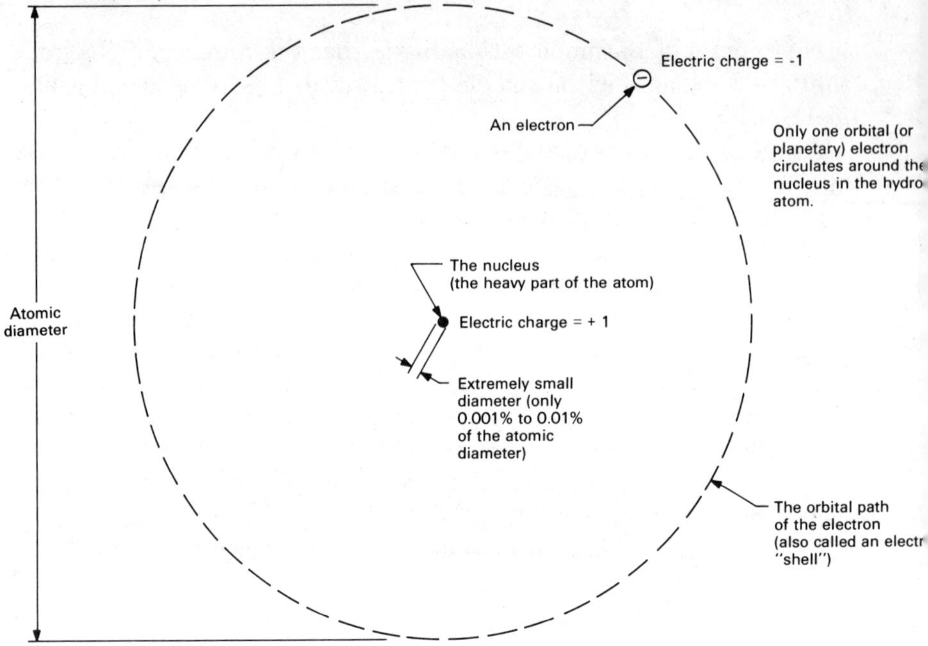

Figure 1.1 The hydrogen atom.

either positive or negative, if extra electrons exist) and therefore ions can be influenced by the presence of an electric field.

If charge is separated by expending some energy and moving some charge from point A to point B, a voltage will exist between point A and point B. This voltage is the restoring force that is trying to reunite the displaced charge. This restoring force is created by the electric field that is established by the voltage that is created between point A and point B by the displaced charge. The magnitude of this restoring electric field (measured as volts per meter) can be expressed as the voltage difference that exists between point A and point B divided by the distance between points A and B. Because the voltage will cause this separated charge to return to its original location, we can make this returning charge flow do useful functions in an electronic circuit. This flow of charge is an electric current.

A voltage is necessary to cause a current to flow; therefore, a current can be thought of as a measure of how rapidly an applied voltage is able to return the charge that was separated, initially, in order to create the voltage. Whenever current flows, an electric "pump" is therefore needed. Something has to exist within a circuit to maintain the voltage, to keep "pumping up the charge" so that it will continue to flow within the circuit. Therefore, we must find ways to take electrons from one place and move them to another place, to "pump them up" within a battery or other power source so that they will create a voltage difference that causes them to return. In Chapter 2 we'll take a look at a number of the ways that this can be done, but we will now return to our discussion of atoms.

The hydrogen atom is the simplest atom: it has only one electron. For the other elements, additional orbital electrons (and also an equal number of additional positive charges, the protons) are added. The large number of elements results by sequentially adding one electron (and also one balancing positive charge in the nucleus) to this simple hydrogen atom structure.

As the number of the protons (and the corresponding number of electrons) increases, the atomic weight of the resulting element increases. In general, the elements with the smaller numbers of protons and electrons tend to be gases. Those elements with larger groupings of electrons and protons are the heavier atoms. In fact, the uranium atom was useful in atomic experiments because of its relatively large size (elements beyond uranium must be produced artificially). Uranium was more easily split and the resulting breaking up of this atom, the atom splitting, released the tremendous energy that is associated with an atomic bomb. Large amounts of energy can be obtained when matter is converted into energy by this atom splitting.

The addition of electrons in the sequencing of the elements is done in a very specific manner. For example, much as the planets circulate around the sun in definite orbits or at definite distances from the sun, the electrons also circulate around the nuclei of the atoms of the elements in definite layers or shells. These orbiting electrons also must maintain specific distances from the nucleus.

As the complexity of the atom increases, only two electrons are allowed in the innermost layer, the electrons closest to the nucleus. After these two electrons exist in the first shell, a second shell starts building. This second shell can contain a maximum of only eight electrons, and then a third shell starts. In general, the basic idea is to add to the innermost electron shells and then add to the next outermost layer of electrons as the elements continue to increase in complexity (although it is actually more complex than this). Outer shells of all elements contain at most eight electrons, and those elements that contain exactly eight are chemically inactive.

Chemical reactions

Chemical reactions are governed by the outermost electrons of an atom. Figure 1.2 indicates the electron shells that move out from the nucleus of an atom. For reference, each electron shell has been given a letter designation. For example, the innermost shell (next to the nucleus) is designated as the K shell, the next outermost is the L shell, then the M shell, and so on to the Q shell.

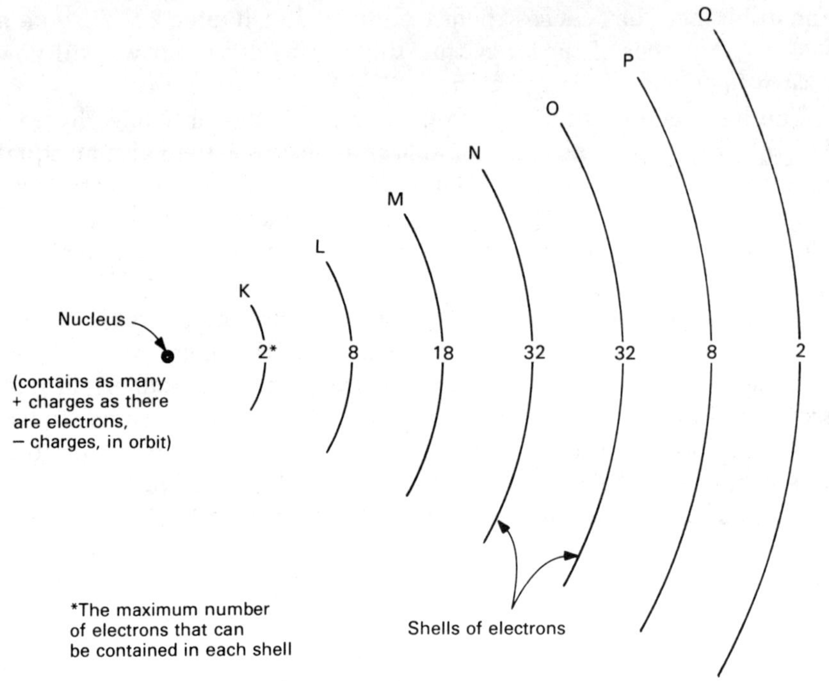

Figure 1.2 The electron shells of the elements.

As electrons are added to make the more complicated or larger atoms of the elements, there will be a varying total number of electrons that occupy the outermost shells of each atom. It is these electrons in the outermost shell of an atom that give the element the characteristics that are needed to form chemical compounds, and also the characteristics that are needed provide the useful semiconductor crystals that are responsible for the present electronic age.

To provide stable chemical compounds, atoms of various elements group together and share their outermost electrons in what chemists call a "covalent" bond. This sharing of electrons between two or more atoms in a chemical compound has a maximum total of eight electrons in the outermost shell around each atom. This magic number of eight is the basis for the various ways that elements combine to form chemical compounds and also is involved in forming crystal structures.

Those elements, like krypton and argon, that naturally have eight electrons in their outermost shells are complete, without any help from any additional electrons that could be contributed by or shared with other elements. Therefore, elements like krypton and argon are called the noble gases because in their outermost shells they already have

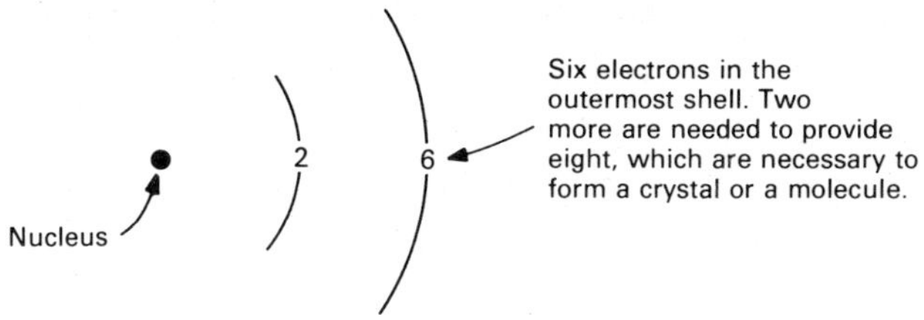

Figure 1.3 The atom of oxygen.

eight electrons. These elements, therefore, are not chemically reactive and are therefore not very useful in providing chemical compounds.

The basic diagram of a more complex atom, an atom of oxygen, is shown in Figure 1.3. A total of eight electrons is associated with this atom. This implies that the innermost shell is filled with two electrons and that the next, outermost, shell contains six electrons. The six electrons in the outermost shell give rise to the chemical characteristics of oxygen.

The chemical symbol for the very familiar molecule of water, H_2O, indicates that a molecule of water is composed of two atoms of hydrogen and one atom of oxygen. The oxygen atom supplies six electrons in its outermost shell, and each hydrogen atom supplies one electron, as shown in Figure 1.4. Adding two hydrogen atoms provides a total of eight shared electrons in one molecule of water. The oxygen atom is shown in the center of this figure. There is a hydrogen atom to the left and to the right, providing the stable system of eight shared electrons. Chemists diagram the sharing of these electrons as H—O—H. The H represents a hydrogen atom and the O, an oxygen atom. The line or bar between these letter symbols indicates the covalent bonding of the electrons.

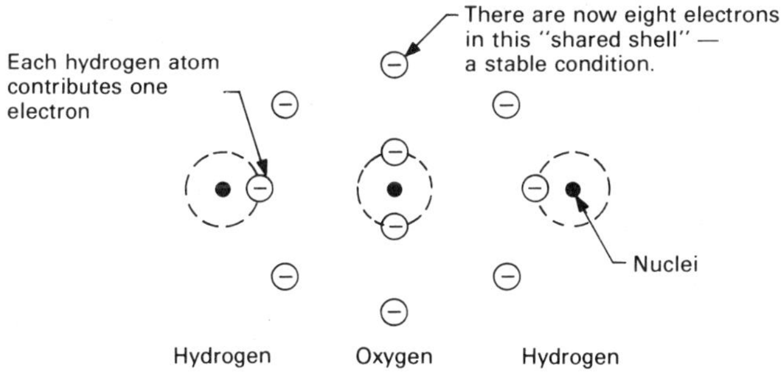

Figure 1.4 A molecule of water.

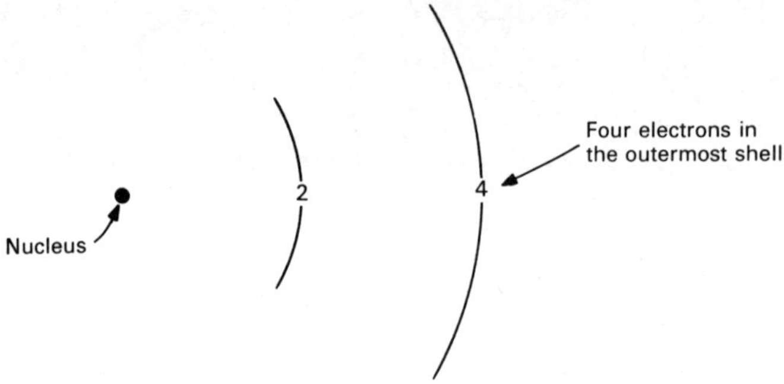

Figure 1.5 The atom of carbon.

The elements that have four electrons in their outermost shells are right at the balance point. They neither have too few, like hydrogen with only one electron, or too many, like oxygen with six valence electrons. In fact, these can be called the "hermaphrodite" elements because they are perfectly happy to combine with themselves.

The first element that has four electrons in its outermost shell is carbon, shown in Figure 1.5. Four is half of the magic number of eight, and this allows carbon to have a lot of different possibilities in forming chemical compounds. All forms of life here on earth, whether plant or animal, are based on carbon. The very broad and important subject of organic chemistry is based on compounds that contain carbon.

As shown in Figure 1.6, carbon can bond with itself. This is a unique thing.

The example is the gas ethane, which has a single covalent bond between two carbon atoms. The six hydrogen atoms each supply an additional electron to form the stable shell of eight electrons around each one of the carbon atoms. So a molecule of ethane can be represented as two atoms of carbon and six atoms of hydrogen. The chemical formula

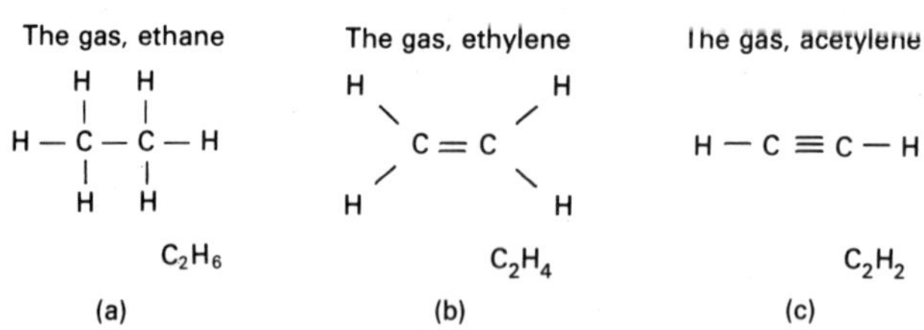

Figure 1.6 Carbon can bond with itself and other atoms. (*a*) Single bonding, (*b*) double bonding, and (*c*) triple bonding.

would be C_2H_6, where the numerical subscripts indicate the total number of each of the atoms that are contained in the molecule.

An example of double bonding with carbon is shown in Figure 1.6*b*. Two adjacent carbon atoms are bonding with each other and each atom is now sharing two electrons. Therefore only two additional hydrogen atoms are needed to provide the stable number of eight valence electrons around each of the carbon atoms. This is the gas ethylene, C_2H_4.

Carbon can also form triple bonds, shown in Figure 1.6*c*, as in the gas acetylene. The triple bond between adjacent carbon atoms now means that only one hydrogen atom is needed for each of the carbon atoms. So the chemical formula for acetylene is C_2H_2.

In all of these examples of carbon bonding, we are showing the simple two-carbon case. Long strings, or chains (and also closed rings), of bonded carbon atoms exist that form the heavier chemical compounds such as the greases and waxes. These dense organic compounds are long-chain bindings of carbon which are linked together and have other atoms also bonded to the carbon atoms.

An interesting thing about carbon is that it can completely bond with itself, as shown in Figure 1.7. There are no atoms involved here other than carbon. Each carbon atom is completely linked up, sharing electrons with other carbon atoms. This now forms a very strong and stable crystal structure. These so-called "valence" crystals (crystals that are formed out of only one element) are extremely tough structures. Very strong electrical forces are involved in this bonding, so these atoms, and even the associated electrons, are very difficult to break loose. This lack of free electrons makes these crystals rather poor conductors of electric current; therefore they are "insulators."

The crystal form of carbon is the high-cost diamond. Because carbon has a small atomic size, diamonds are extremely hard materials. Diamonds, for example, can scratch all other materials because the stronger electrical bonding between the atoms of a diamond allows it to break up the atomic bonding of other materials.

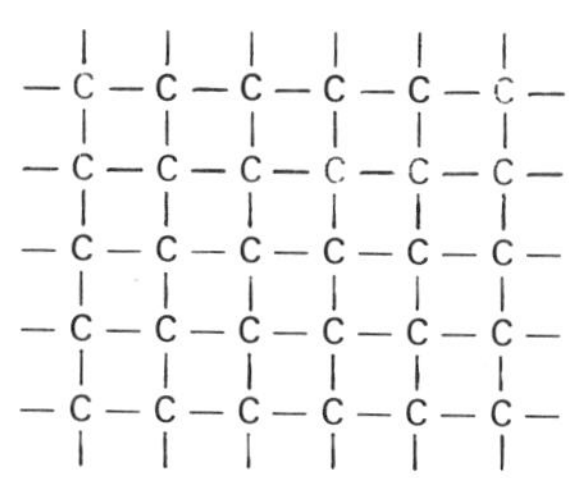

Figure 1.7 Carbon bonding with itself.

With these outermost electrons completely tied up in covalent bonding, a diamond will not conduct an electric current and therefore forms a solid insulator. In fact, a diamond remains an insulator even at relatively high temperatures.

The semiconductors

Most of the good conductors, such as the metals copper, silver, and gold, form solids in which the individual atoms of these metals are surrounded by a loosely held "sea" of their accompanying electrons. When a voltage is applied across a metal, this large number of available (free or mobile) electrons is responsible for the large current that flows.

Solid crystal structures are basic to semiconductors. We will be emphasizing semiconductors because of their importance in modern electronics. Semiconductors form valence crystals that have a crystalline structure like a diamond. Most of the electrons within semiconductors are relatively tightly held in position; they are not free to move around within the crystal structure. Therefore, when a voltage is applied across a semiconductor, there are only a few mobile electrons that are available to carry a current. These free electrons have been provided by the thermal energy that exists at room temperature, which causes a few covalent bonds to be randomly broken within the crystal structure.

The next elements in the sequence after carbon that have four electrons in their outermost shell are silicon and germanium. These are extremely useful for electronics because their larger atomic size doesn't give rise to such intense bonding forces on the electrons as exist in a diamond. Therefore, thermal energy is able to break or rupture a few covalent bonds and thereby free occasional electrons (and "holes," which we will soon describe) that can be made use of to allow a current to flow through the solid crystal structure. (Special materials, called thermistors, are also available that make use of this reduced resistance at elevated temperature characteristic to provide a temperature-sensing electrical component.)

The elements germanium and silicon are very similar to carbon. They bond with themselves and form relatively strong crystal structures. Germanium and silicon are not good insulators, like diamonds, and yet they do not have the large supply of free electrons that exist with metals. This causes germanium and silicon to be neither good insulators nor good conductors. Therefore, they are called "semiconductors." Today, the emphasis is on silicon; so we will restrict our discussions to this popular semiconductor.

The electron shells for an atom of silicon are shown in Figure 1.8. Notice that the first shell is filled up with two electrons, the second

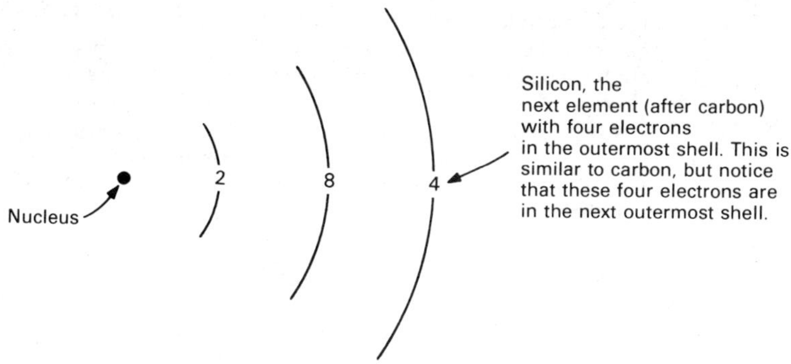

Figure 1.8 The atom of silicon.

shell with eight, and four electrons remain in the outermost shell, for a total of fourteen electrons.

This atom is similar to an atom of carbon because four electrons are available in the outermost shell of each of these elements. Therefore, in a manner identical to carbon, silicon can bond with itself to form a very stable crystal that has a crystalline structure like a diamond. Most of the electrons within the crystal structure of silicon are relatively tightly held in position, so they are not free to move around. Therefore, when a voltage is applied across a piece of silicon, there are only a relatively few electrons available to carry a current. These free electrons are provided by the thermal energy that exists at room temperature, which causes a few covalent bonds to be randomly broken within the crystal structure. As the temperature is increased, the resistance of silicon decreases. It is this single crystal of silicon that is the starting material in fabricating semiconductors.

Figure 1.9 illustrates the sharing of electrons in the covalent bonding of silicon atoms. The circle with the silicon symbol (Si) in the middle represents the nucleus of a silicon atom. The minus sign with the circle around it represents an electron. The solid line that connects an electron back to an atom indicates that this electron can be thought of as being associated with that particular atom of silicon. A dashed line indicates the sharing of an electron by a neighboring silicon atom. So, one atom supplies an electron, and an adjacent atom shares this electron. This covalent bonding of the silicon atoms exists in three dimensions to form a solid structure as shown in Figure 1.10. In this way, a stable set of eight electrons surrounds each of the silicon atoms, and a very stable, well-ordered crystal structure results.

Silicon is useful in electronics because it can be modified by exchanging a relatively small number of the atoms within the crystal structure with atoms of a different element. This modified or contaminated

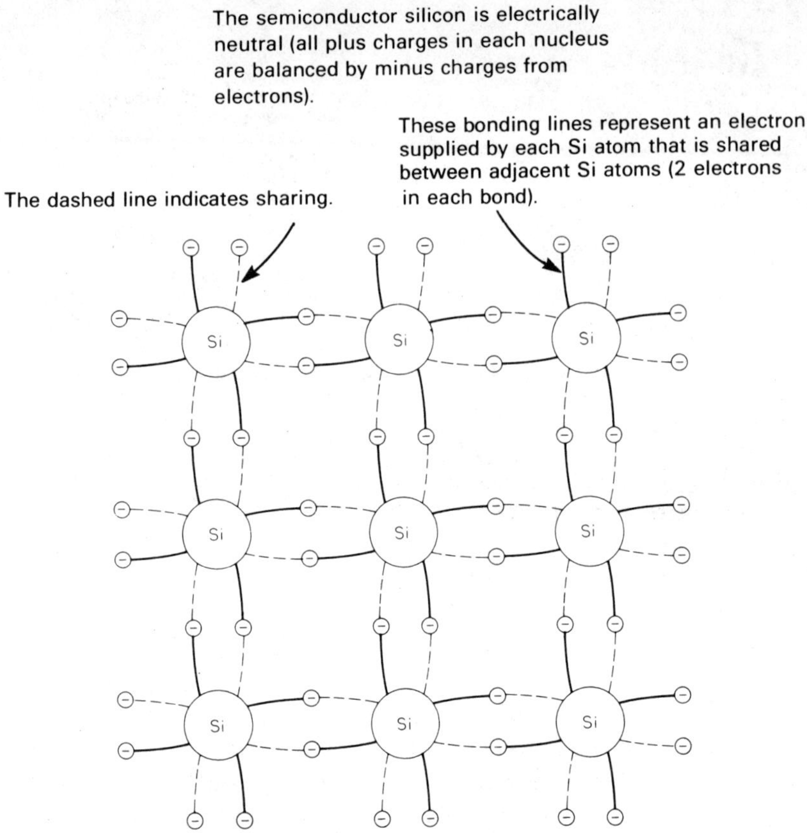

Figure 1.9 A structure formed only with silicon.

silicon is called "doped silicon." The process of doping is the occasional substitution of a different atom for a silicon atom in the solid structure of a silicon crystal.

There are two very useful possibilities when doping atoms are substituted into the crystal structure of silicon. A doping atom with only three electrons in its outermost shell will create a place for an additional electron, a "hole." This hole is then available to conduct an electrical current. Or, alternatively, a doping atom with five electrons in its outermost shell will provide *a spare electron* within the crystal structure of silicon that can be easily stripped away: a free electron. This free electron is then available to conduct an electrical current. The conductivity of silicon will increase with either P-type or N-type doping.

This gives rise to what is known as P type for those doping elements that have less than four valence electrons because they will provide a

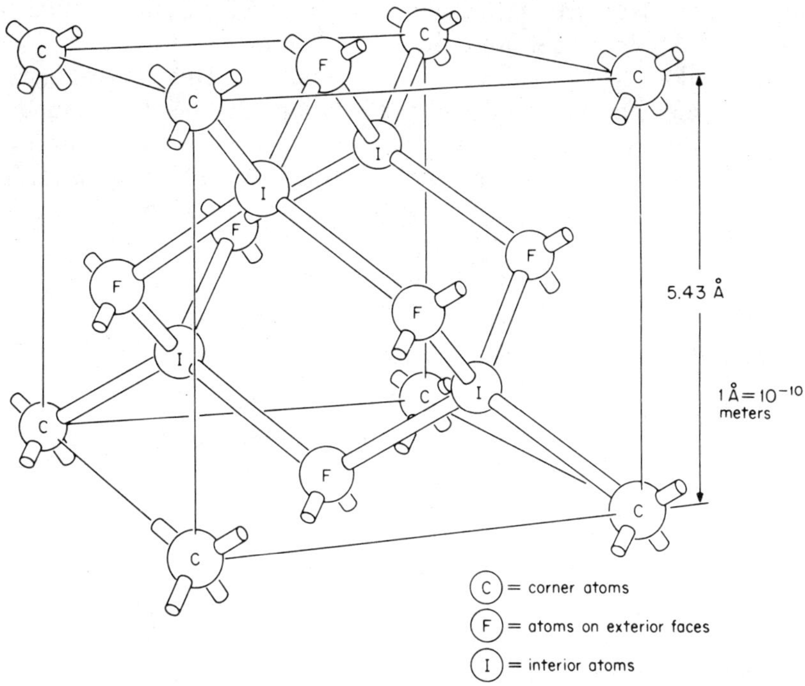

Figure 1.10 Three-dimensional bonding of silicon to form a solid crystal.

place for an electron, a hole. These holes act essentially the same as if they were positive charges, and therefore the P-type designation is used.

A designation of N type is used for those doping elements that introduce a spare electron (a negative charge) in the crystal structure of silicon.

A bar of pure silicon material, in a nice crystal order, would not be a very good conductor. This means that if a voltage were applied between two wires that are attached to the silicon, not much current would flow. But, if this semiconductor material were doped with either P-type or N-type dopants, current would flow. Holes would provide temporary resting places for conducting electrons as they moved through a P-type-doped crystal, or the free electrons would allow current to flow for an N-type-doped crystal. A bar of doped silicon therefore becomes a resistor, an electrical component that conducts more current than an insulator but less current than a conductor. Doped regions of silicon are used to form resistors within integrated circuits.

When a region that has been doped P type is bordering a region that has been doped N type, the very useful PN junction (a diode) is formed (diodes are discussed in Chapter 4). Diodes are also basic to the operation of bipolar transistors (the solid-state amplifying devices, also discussed in Chapter 4) that are used in electronics.

We will take a look at silicon again in Chapter 4. This general background on elements, atoms, covalent bonding, chemical compounds, semiconductors, and doping provides a framework with which to understand how a solid piece of single-crystal silicon can be converted into many useful electronic things such as diodes, transistors, resistors, and capacitors. In modern integrated circuit (IC) electronics, essentially all of the circuit elements are made out of silicon.

Dealing with Very Large and Very Small Numbers

In the calculations that are used in electronics, numbers are both extremely large and extremely small. Very large numbers, like 1000, 10,000, 100,000, or 1,000,000, require writing down a large number of digits. The notation technique that scientists use to handle very large numbers is shown in Figure 1.11. This simplifies the problem of excessive writing, makes calculations easier to handle, and reduces errors that are often made in calculations.

In scientific notation, large numbers are expressed as a small number ranging between 1 and 9 (like 2.98 or 4.53) and then a separate "scaling factor" is used to show how big the number is. Powers (exponents) of 10 handle the scaling factor, as shown in Figure 1.11. A scaling factor of 10 to the first power, 10^1, is 10. A scaling factor of 10^2

Numbers from 1 to 9 are easily handled in calculations.
To make calculations easier, a very large number can be expressed as a number from 1 to 9, followed by 10, 10^2, 10^3, etc., because:

$10^0 = 1$
$10^1 = 10$
$10^2 = 100$
$10^3 = 1000$

Therefore:

120. V = 1.2×10^2 V (Check: $1.2 \times 100 = 120$)

First, move the decimal point two places to the left to express the number in the range 1 to 9. The number of places moved to the left (2) then becomes the exponent of 10.

And,

490000. V = 4.0×10^5 V
54321

Figure 1.11 The scientific way to write very large numbers.

(10 × 10) indicates 100 times, and 10^3 (10 × 10 × 10) indicates 1000 times. (10^0 has been defined to be equal to 1.)

For example, 120 volts (V) can be expressed in scientific notation as 1.2×10^2 as shown in Figure 1.11. This can be thought of as moving the decimal point from 120.0 to 1.20. The decimal point was moved to the left two places; therefore 2 is the exponent on 10.

The second example in this figure shows a much larger number: 490,000 V, a very large voltage. Rather than writing this as 49 with four trailing zeros, it is certainly easier (and less prone to errors) to write 4.9 (that is the number between 1 and 9) and then indicate how big the number is as an exponent on 10. In this example the decimal point had to be moved five places to the left, so the exponent is 5. The notation 4.9×10^5 is an easier way of writing 490,000.

Figure 1.12 shows how this same concept is used to express very small numbers. Many small numbers that may be only a millionth or a millionth millionth part of 1 must be written down and handled in calculations. Again, exponents of 10 are used, but these are negative exponents. The expression 10^0 is 1, 10^{-1} is 0.1, 10^{-2} is 0.01, and 10^{-3} is 0.001. For example, 0.028 V is written as 2.8×10^{-2}. In this example the decimal point was moved to the right two places in the original expression to arrive at the number to place the numerical part of the number in the range of 1 to 9; the exponent is −2. In scientific notation, moving the exponent of a number to the left creates a positive exponent; moving the exponent to the right creates a negative exponent. This can be remembered because moving the exponent to the right is making the number larger. Therefore a negative exponent of 10 must be used to bring it back down to where it started.

Use the same idea, express the number in the range 1 to 9, but now follow with 10^{-1}, 10^{-2}, 10^{-3}, etc., because:

$$10^0 = 1.0$$
$$10^{-1} = 0.1$$
$$10^{-2} = 0.01$$
$$10^{-3} = 0.001$$

Therefore:

$0.028 \text{ V} = 2.8 \times 10^{-2} \text{ V}$

1 2

You now move the decimal point to the right to express the number in the range 1 to 9. This number of places moved now becomes a negative exponent of 10.

And:

$0.000\,056 \text{ V} = 5.6 \times 10^{-5} \text{ V}$

123 45

Figure 1.12 The scientific way to write very small numbers.

Another example, 0.000056, is shown in Figure 1.12. This number is difficult to write down. In this case, the decimal point must be moved to the right five places. It is much easier to refer to this number as 5.6×10^{-5} rather than to write down all of those zeros.

In electronics there are very large values of resistance, very low values of capacitance, and very low values of current flow. This shorthand way to write down these wide ranging numbers is very convenient.

Handling the Calculations that Aid Understanding

When doing calculations with numbers that are very large and very small, you can quickly see the benefit of using scientific notation. For example, the problem of multiplying a couple of the numbers from the previous figures is shown in Figure 1.13, 490,000 V multiplied by 0.000056 V.

Both of these numbers are first expressed in scientific form as 4.9×10^5 V and 5.6×10^{-5} V. The numerical parts, 4.9 and 5.6, are multiplied, and then the scaling factors are multiplied. The numerical part of this product (4.9×5.6) can be handled on a standard four-function calculator.

To multiply the scaling factors, simply add the exponents of 10. Adding the exponents +5 and −5 results in 0, and 10^0 is the same as 1. Multiplying any number by 1 does not change the number; so 10^0 can be simply dropped. The answer therefore is 27.4.

Both of these numbers are voltages. Volts times volts provides $(\text{volts})^2$, or V^2. It is very important to carry the units in all calculations. When the units properly cancel or combine, it indicates that you are using the correct equation. In this particular case, the product is expressed as 27.4 V^2, or 2.74×10 V^2.

Instead of multiplying numbers, let's take a look at an example in which these numbers are divided, as shown in Figure 1.14. For simplicity, the same numbers will be used, 490,000 V ÷ 0.000056 V. After

Multiplying the number from the previous figures:

(490,000V) (0.000,056 V) = ?

First, express both numbers in scientific form:

$(4.9 \times 10^5 \text{ V})(5.6 \times 1C^{-5} \text{ V})$

or $(4.9)(5.6) \times [(10^5)(10^{-5})]\ V^2$ *

or $27.4 \times 10^0\ V^2$

$27.4\ V^2$

When multiplying the powers of 10, algebraically add.
Here: + 5 − 5 = 0 and $10^0 = 1$

*Note: $V \times V = V^2$
because the units also multiply.

Figure 1.13 Multiplying very large and very small numbers.

Dividing the numbers from the previous figure:

$$\frac{490{,}000\ \text{V}}{0.000{,}056\ \text{V}} = ?$$

First, express each number in scientific form:

$$\frac{490{,}000\ \text{V}}{0.000{,}056\ \text{V}} = \frac{4.9 \times 10^{5}\ \text{V}}{5.6 \times 10^{-5}\ \text{V}}$$

Note: The sign of the exponent is changed when it is moved up into the numerator.

or $$\left(\frac{4.9}{5.6}\right) \times \left(\frac{10^{5}}{10^{-5}}\right) = \left(\frac{4.9}{5.6}\right) \times [(10^{5})\,(10^{5})]$$

Exponents add when multiplying.

$$\left(\frac{4.9}{5.6}\right) \times 10^{10} = 0.88 \times 10^{10} = 8.8 \times 10^{9} \text{ (a dimensionless ratio)*}$$

*Note: V/V = 1; The units cancel.

Figure 1.14 Dividing very large and very small numbers.

these numbers are written in scientific form, the numerical part of the number is separated from the scaling part to give 4.9 ÷ 5.6.

To divide these scaling factors of 10, *the exponent of 10 that is in the denominator is moved up into the numerator by simply changing the sign on the exponent of that factor of 10.* This provides $10^5 \times 10^5$. The exponents are simply added to accomplish the multiplication. So, the power (or the scaling factor) of this quotient will be 10^{10}, a very large number (this is approximately the number of stars that exist in the observable sky when the best telescope is used).

When the numerical parts are divided (4.9 ÷ 5.6), the quotient is a little less than 1 (0.88). The scaling factor of this quotient is 10^{10}. Again, this quotient will be expressed in scientific form, so the decimal point is moved one place to the right. This makes the numerical part of the quotient larger; therefore the scaling exponent of 10 (also 10) must be reduced by 1 to bring it down to 9. So, the answer for this indicated division is 8.8×10^9. Notice that volts were divided by volts, so the volts cancel. This quotient is now a dimensionless ratio. One voltage is compared with another voltage.

With modern calculators, multiplications and divisions are easily done. Usually, the scaling factor powers of 10 can be handled in your head because this involves adding and subtracting relatively small numbers. Therefore, a four-function calculator can be used to handle these calculations. If you do happen to have a scientific calculator, you can make use of the ability to directly enter the exponent along with each number. Then you won't have to handle the exponents by hand calculations. If you would like to pursue the study of electronics, the investment in a scientific calculator is certainly worthwhile.

We will use an example of playing with calculations to show what can be done when the units are carried along with the numbers. For example, what happens if we tried to answer the question: How many seconds exist in 1 year? Well, we start, as shown in Figure 1.15, by knowing that there are 60 seconds (s) in 1 minute (min). We write 60 s over (divided by) 1 min. Because these are the same thing (60 s are the same as 1 min), this expression has a value of 1. Now, we want the number of seconds in a year, so we don't want minutes in the denominator. The minutes can be canceled out by remembering that there are 60 min in an hour, so we multiply our starting expression by 60 min over 1 hour (h). Now the minutes cancel, leaving the number of seconds in 1 h. But, we still want a year, so let's work our way to larger time units in the denominator. We know that there are 24 h in a day; so multiplying by this factor, we can cancel the hours and obtain the number of seconds in 1 day. But, a day is still not a year, so we have to finally put in the last factor that there are 365 days in 1 year. Now the days cancel, leaving seconds in the numerator (that's what we want) and years in the denominator (this is also what we want). We simply have to multiply all of these conversion factors: $60 \times 60 \times 24 \times 365$.

In the figure, we show an example in which we are expressing all of these factors in scientific form, separately calculating the numerical part, and then handling the scaling factors as exponents on 10. We finally express this overall product in scientific form to give the final answer: there are 3.15×10^7 s in 1 year (a relatively large number of seconds).

In electronic engineering, the "metric prefixes" are used as scaling factors. This then provides a word description for the scaling factor. It is much easier to use a word description than to say, "Three point one

The number of seconds in 1 year:

$$\frac{60\text{ s}}{1\text{ min}} \times \frac{60\text{ min}}{1\text{ h}} \times \frac{24\text{ h}}{1\text{ day}} \times \frac{365\text{ days}}{1\text{ yr}} = (60)(60)(24)(365)\text{ s/yr}$$

$$= (6 \times 10)(6 \times 10)(2.4 \times 10)(3.65 \times 10^2)\text{ s/yr}$$

$$= (6)(6)(2.4)(3.65) \times 10^5\text{ s/yr}$$

$$= 315 \times 10^5\text{ s/yr}$$

$$= 3.15 \times 10^7\text{ s/yr}$$

Start here with the number of seconds that exist in a larger unit of time.

Keep canceling the time units you don't want by moving toward the final unit (year) you do want. Each of these multiplying factors has a value of 1, so it has no effect other than changing the units.

Figure 1.15 Calculating the number of seconds in a year.

The metric prefixes for the positive exponents of 10 are:

10^3 = kilo (k)
10^6 = mega (M)

Therefore:

1×10^3 V = 1 kV

and

3.2×10^6 V = 3.2 MV

The sum of these exponents is still 5.

and

4.95×10^5 V = $4.95 \times 10^2 \times 10^3$ = 495 kV rather than 0.495 MV

Figure 1.16 The engineering way to write very large numbers.

five times ten to the seventh" (3.15×10^7). The standard metric prefixes are used, whether the numbers refer to lengths or weights or (in the electronic case) ohms, volts, or amperes.

For example, as shown in Figure 1.16, the metric prefix for 10^3 is "kilo" (k). You probably have heard of kilograms (kg) or kilometers (km); that is a thousand grams or a thousand meters. In engineering, 1×10^3 V would be called 1 kilovolt (kV).

The metric prefix for 10^6 is "mega" (M); that is a million. As an example, 3.2×10^6 V would be referred to as 3.2 megavolts (MV) (3,200,000 V).

The last example in Figure 1.16 is 4.95×10^5 V, but 10^5 is not one of the standard metric prefixes. If this scaling factor were brought down to 10^3, it would be kilo. This can be done by expressing 10^5 as $10^2 \times 10^3$. The 10^2 is moved into the numerical part (4.95) as 495. Therefore, 4.95×10^5 V becomes 495 kV. The next larger metric prefix could also be used as 0.495 MV. This designation would be used if the other units that are being discussed happened to be megavolts.

Small numbers use the metric prefixes shown in Figure 1.17. In electronics, there are usually more small numbers than large numbers.

Use metric prefixes for the negative exponents of 10 as:

10^{-3} = milli (m)
10^{-6} = micro (μ)
10^{-9} = nano (n)
10^{-12} = pico (p)

Therefore:

2.8×10^{-2} V = $\underbrace{2.8 \times 10}$ ($\times 10^{-3}$) = 28 mV

5.6×10^{-5} V = $\underbrace{5.6 \times 10}$ ($\times 10^{-6}$) = 56 μV

9.3×10^{-7} V = $\underbrace{9.3 \times 10^2}$ ($\times 10^{-9}$) = 930 nV

Figure 1.17 Metric prefixes for small numbers.

Whereas kilo and mega are generally all that are needed for the large numbers, many more metric prefixes are used for small numbers. Therefore, metric prefixes are used for 10^{-3}, 10^{-6}, 10^{-9}, and 10^{-12} (metric prefixes only exist for each factor of one thousand). So, 10^{-3} in metric is "milli" and is abbreviated as m (you have probably heard of millimeters in cigarette ads). Milliamperes (mA) and millivolts (mV) are used in electronics. A millionth part, 10^{-6}, is called "micro," and the abbreviation is the Greek letter μ. Even smaller factors are 10^{-9}, "nano" (n) and 10^{-12}, "pico" (p).

For example, the scientific number 2.8×10^{-2} V would be expressed as 28 mV. Similarly, 5.6×10^{-5} V would be called 56 μV. These are certainly easier to say than if these numbers were expressed in scientific form. This metric format is a way to more easily talk about small and large numbers.

With this background on calculations, the strangeness of the numbers that are encountered in electronics can be eliminated. These scientific and metric number representations will be used in calculations throughout the discussions in this book so that you can get a feeling for how this game is played. Unfortunately, there is a lot of mathematics associated with electronic design, but if you are not going to become a circuit designer, all you need is a speaking acquaintance, a recognition of what the words mean so you can understand what is going on.

The Displays of Three-, Three-and-a-Half-, and Four-Digit Voltmeters

Display devices make use of a strange sounding way to describe the number of digits of numerical data that they can indicate. For example, the digital multimeter that is recommended to accompany this book is called a "three-and-a-half-digit meter." The obvious question is, "What is this half digit?" This sounds confusing, and the answer lies in the way that the electronics for each digit is provided.

As in counting, an overflow or carry out of a lower digit to the next-most-significant digit is provided. The trick of the half digit is to make use of this overflow out of the most significant digit to increase the range of numbers that are displayed. All of the expense of the next-higher-digit electronics is not needed. This low-cost, easily added extra digit displays either nothing (representing zero) or a 1. As shown in Figure 1.18, this relatively simple half-digit feature extends the range of the three-digit display of Figure 1.18*a* from 1000 steps (000 through 999) to 2000 steps (0000 to 1999) that are realized when the half digit is added, as shown in the three-and-a-half-digit display of Figure 1.18*b*. If all of the electronics for the extra digit were provided, the four-digit

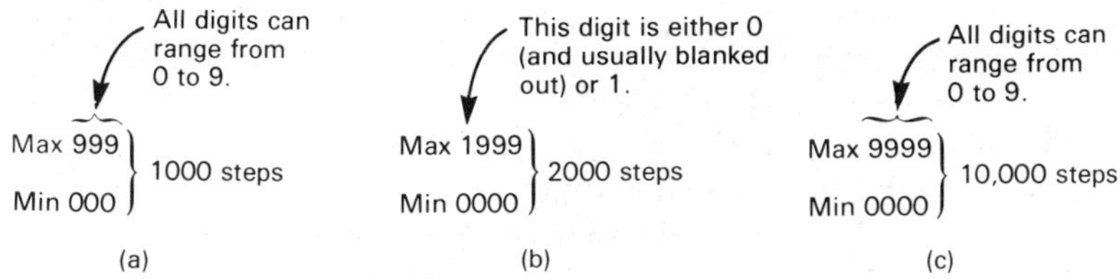

Figure 1.18 The resolution of (*a*) 3-, (*b*) 3½-, and (*c*) 4-digit voltmeters.

display of Figure 1.18*c* would result, and this would increase the display range to 10,000 steps (0000 to 9999).

Another confusing thing about displays is the full-scale range. For example, the digital multimeter used with this book makes use of full-scale ranges that are 2, 20, 200, etc.; yet the maximum display indication is only 1.999, 19.99, 199.9, respectively. The display range of the meter indicates how to interpret the scaling factor of the displayed digits; the display will not provide the actual full scale numbers of 2, 20, 200, etc.

What Is an Electronic Circuit?

An "electronic circuit" uses an energy source, such as a battery or a power supply, and some wire to hook up various circuit components that include active devices (transistors) in such a way that a useful electronic function will be obtained. There are many different types of electronic circuit components. These range from the passive components (which are also used in electric circuits), such as resistors, capacitors, and inductors, to the active components, such as transistors (that are unique to electronic circuits and provide voltage gain). Chapter 4 will describe some of the passive components that are used in electronic circuits.

Whenever there are no transistors or other amplifying devices in a circuit, it is called an "electric circuit," not an electronic circuit. In homes, the lighting system and the control of the heating system are electric circuits in contrast with radios, calculators, and home computers, which use electronic circuits.

Getting Familiar with the Lab Kit

You will find that the 200-in-One Electronic Project Kit (Radio Shack Cat. No. 28-265) is a very convenient way to learn about electronics. Within this kit are a lot of the basic circuit components that are used in making electronic circuits. The springs associated with each one of these components make it very easy to use the wires provided to hook up these components with each other to make circuits.

The basic idea of this 200-in-One kit is to make circuits that make noise and flash lights to stimulate children to become interested in electronics. The approach that we will take is to add a very precise meter for use with this lab kit. By using both of these, and some specific electronic circuit demonstrations, you can learn much more about electronics than can be accomplished with only a book or a series of lectures.

I would encourage you to take the manual that comes with the lab kit and to experiment with any of the 200 circuits that you find interesting. A laboratory experience is necessary to make electronics seem real because electronics deals with many concepts that are very strange and are outside the realm of normal experience. For example, how can you feel good about current unless you measure current and unless you have the ability to set up a circuit and to predict, in advance, what the current flow should be and then measure it to see if you are right? So, after you complete this book, go on and build some of these fun things and also read the manual that comes with this kit so that you can expand your knowledge of electronics. Spend the additional time and get the additional information. This should be a pleasurable thing; experimenting with electronics is an interesting hobby for many people.

With the combination of this kit and the meter, you'll find that you can go on and do fundamental measurements on any specific electronic components that you may be working with. For example, if you happen to be working with diodes, you can make some measurements on them to get a better idea of the diodes that you are dealing with—or the resistors, the capacitors, or any other electronic components that you deal with or are interested in. The information that you can learn from this book will provide enough background so that you will be able to use your imagination to set up the necessary experiments.

We will use this lab kit, shown in Figure 1.19*a*, and the digital multimeter (DMM), shown in Figure 1.19*b*, to provide a neat assemblage of basic electronic components and measuring capability that will assist your learning.

If you take a look at the lab kit you will see in the upper left-hand corner that there are 20 resistors. Adjacent to this are two silicon diodes. Transformers (an input transformer and an output transformer) and some batteries are in the upper right-hand corner. Two digital integrated circuits, a dual *JK* flip-flop, and a quad two-input NAND gate are along the center right. There is a relay, and many other electronic components are also available.

Across the middle of the kit you will see that there are four transistors. The two transistors to the left are PNP transistors and the two on the right are NPN (we will discuss transistors in Chapter 4). Just to the left of that group of transistors is a germanium diode that is labeled with the abbreviation "Ge." To the left of the germanium diode is an

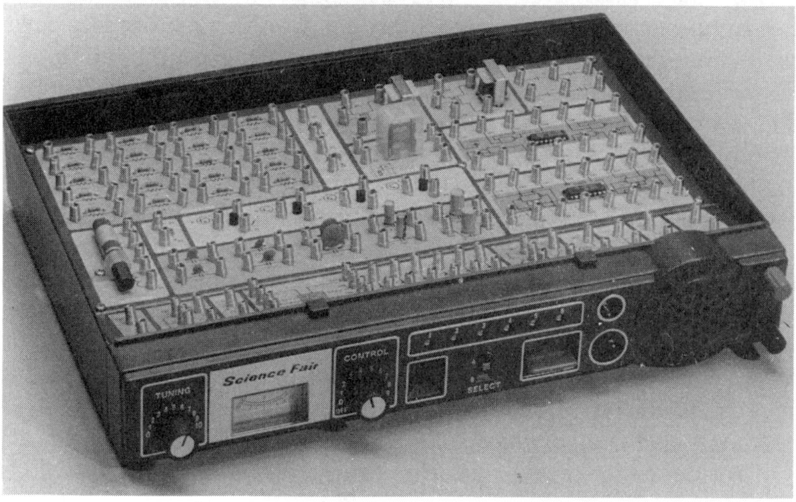

Figure 1.19 The equipment that we will use to perform experiments. (*a*) The 200-in-One electronic Project Lab Kit and (*b*) the digital multimeter (both from Radio Shack).

antenna coil. Immediately under the transistors, and still near the center of the kit, is an array of 10 capacitors. These capacitors range from small values of capacitance, 100 picofarads (pF), in the upper left corner to the extreme right edge where you will see the larger-valued capacitors, 100 microfarads (μF) each. We will have more to say about capacitors in Chapter 4.

Across the lower part of this kit you will notice, to the left, a tuning capacitor. That's the thing labeled "Tuning." The symbol is a capacitor with an arrow through it. A knob on the front panel on the extreme left side is called the "tuning control," and this knob is used to vary the electrical value of this capacitor.

To the right of the tuning capacitor is a thing labeled "Control." A knob on the front panel is also labeled "Control." This control is a potentiometer and is used, for example, to control the audio volume in a radio or a television set. A switch is mounted on the back of this potentiometer. If you turn the control knob to the maximum counterclockwise direction, it will point to OFF and you will hear a click. Rotating the control knob a little bit in the clockwise direction, you will hear another click. This click on and click off is the switch function that is typically provided on the back end of a volume control. This switch is often used as the ON/OFF control for a radio.

On the front panel of this lab kit is also a meter. This meter is not as precise, as nicely calibrated, or as easy to read as the digital multimeter that we will be using. To the right of this meter is a light-emitting diode (LED) digital display. This is a peculiar diode that generates light when current is passed through it. LEDs require less power than incandescent lamps and are also more rugged and reliable, so LEDs are often used in displays for electronic equipment.

There are six AA cells in a lower compartment just under the upper right-hand corner. We will not be dealing with high voltages; nine volts is the highest. It is very hard to electrocute yourself while experimenting with this lab kit. This is a convenient way to learn about electronics without the dangers of plugging anything into a wall outlet.

This lab kit uses an easy way to interconnect the electronic components that consist of a number of wires of various lengths. Contact springs are provided for all of the components, as shown in Figure 1.20*a*. To connect a wire to a contact spring, the top of the spring is pushed sideways, as shown in Figure 1.20*b*, and then the end of the wire is slipped between the opened spring coils. When the spring is released, it will return to its normal position, and the end of the wire will be firmly contacted, as shown in Figure 1.20*c* (a slight tug on the wire is recommended to verify that a good electrical contact is provided by the spring).

Multiple wires can be attached to the same contact spring, as shown in Figure 1.21. Up to three wires are easily contacted if the wires are spaced around the spring so that previous wires don't drop out as the next one is added.

A problem can exist. The wire ends, which are attached to closely spaced springs, can inadvertently touch or short out, as shown in Figure 1.22. Always check to ensure that this undesired connection is not happening as a circuit is being hooked up.

As you go through this book and study each component, you'll use

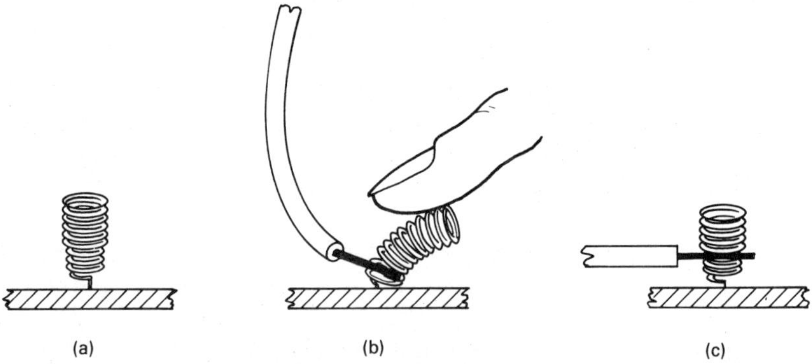

Figure 1.20 Connecting wires between kit components. (*a*) Contacting springs used in lab kit, (*b*) bend spring over just enough to allow slipping the end of a wire between the coils of the spring, and (*c*) let spring return with wire stuck in the coils of the spring. (Pull gently on the wire to check the contact.)

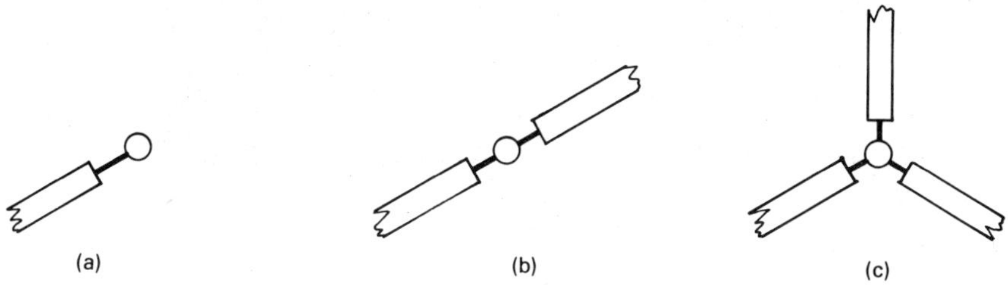

Figure 1.21 Connecting more than one wire to a spring terminal. (*a*) One wire, (*b*) two wires, (*c*) three wires.

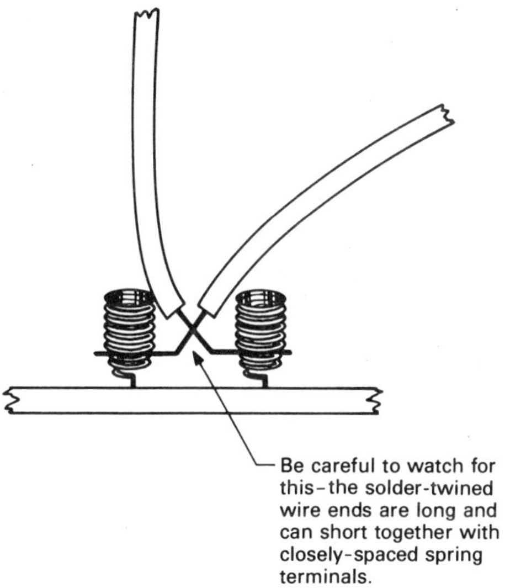

Figure 1.22 Don't let hookup wires short out.

this lab kit and make some basic measurements and observations about each component as you are learning about it. You will therefore have a hands-on opportunity to see many electronic components, hook them up, make useful circuits, make meaningful measurements, and in this way greatly increase your understanding of electronics.

Now, with these preliminaries out of the way, we'll start our investigation of electronics by first considering voltage.

Chapter

2

Voltage: The Pressure of the Electrons

Voltages are most commonly associated with batteries. For example, many people are familiar with the standard 1.5 V that is provided by flashlight cells or the 12 V supplied by the battery in an automobile. In addition, there is the 110 v that is supplied to homes (v means ac volts; a means ac amperes). So, most people are somewhat familiar with a number of voltages, ranging from relatively small voltages (1.5 v) to larger voltages (110 v).

In electronic circuits, *voltage provides the electric force* (the equivalent of water pressure in a hydraulic system) *that causes the electrons to move around, to flow through a circuit.* The movement of these electrons constitutes the current that flows. So, a source of voltage is necessary to make any circuit "come to life."

Voltage sources are provided in many ways. For example, homes have electrical outlets on the walls, and various appliances can be plugged into these outlets to obtain 110 v. Electronic systems are basically designed to be either plugged into an outlet where power is provided or to operate from batteries or cells, as in a portable system like an electronic wristwatch.

What is considered a large voltage? If the voltage is large enough to kill someone, it is generally considered a large voltage. In electronics, up to about 30 V is considered low voltage. Many linear electronic systems make use of +15-V and −15-V power supplies. Digital systems typically operate from a single +5-V power supply.

The 110 v that is available in your home can cause electrocution. If you are not careful and your hands contact the two terminals of a wall electrical outlet, the voltage will cause current to flow through your body. This current flow can interrupt the natural functioning of your heart and can cause death. This electrocution problem is more dangerous if your body is damp or wet with salty perspiration (and therefore a good electrical conductor), or if you reach out of a bathtub to tune in a radio that is also in the bathroom or to turn on or turn off an electric heater or a lamp that is used in the bathroom. Always be very careful. Don't touch any electrical appliances—not even a light switch—if you are sitting in a bathtub or standing in a shower. Electrocution is a prevalent problem. In the home it is of greater concern in the kitchen and bathroom areas because plumbing is exposed, and these pipes that bring in water are typically a good ground—that is, the water pipe is electrically the same as one side (the "cold" or "ground" side) of the 110-v wall outlet. The metal case of a defective appliance can bring the "hot" side of the 110-v line in contact with your hand. If some other part of your body is touching a water pipe, you will then complete a circuit and electricity can flow through you and possibly create electrocution.

I remember one time when I was in a neighbor's garage. I was walking across the cement floor with my shoes on and just happened to touch the top of a large, chest-type freezer. I noticed a little tingling sensation. So, I asked him how long has this been going on. "Oh, it's always been that way," he answered. I said, "Just a minute," got a 100-watt (W) light bulb in one of those portable clamp-on lamps and a couple of clip leads (wires with contacting clips on the ends). I attached one side of the plug for the light to his cold water pipe and attached a clip lead from the other plug contact over to some metal on his freezer. When the 100-W light bulb lit brilliantly, it got his attention. That demonstrated that the hot side of the 110-v line was contacting all of the metal on that freezer. If anybody ever happened to have one hand on the freezer and then reached over with the other hand and touched a cold water pipe, under the right conditions they could die from electrocution!

This dramatic light-bulb voltage detector demonstration was more believable than simply showing him the reading on a voltmeter because voltmeter readings don't mean anything to most people. Seeing a 100-W light bulb glow brilliantly brought the point across. Larger problems exist with the 220 v that is typically brought into electric ranges, electric water heaters, and electric dryers. This 220 v is more dangerous because it's twice the voltage.

In this course, a battery-powered meter and a battery-powered laboratory kit are used. We are not encouraging use of the meter anywhere around the 110-v wall outlets. The intention is to teach something about electronics without risking lives.

In general, everyone considers voltages larger than 200 V to be high voltages. So, 200 V is high, and 10,000 V and larger are very high voltages. Very high voltages do exist in many common products. For example, every automobile has a special coil that is used to raise the 12-V battery voltage up to 10,000 to 50,000 V to "fire" the spark plugs. It's the spark that results from this high voltage across the gap of each spark plug that ignites the mixture within the cylinders of an automobile engine. You can get a good shock from the spark plug wires in an automobile. Also, television sets apply very high voltages to make the picture tubes operate, and many people hurt themselves by reaching in the back of a television while it is plugged in. Television sets have a safety interlock, so when you take the back panel off of a television set, this interlock automatically disconnects the 110-v wires so you can not power the TV set while the electronics are exposed. But, people build "cheater cords" that are used to bypass the safety interlock and apply 110 v to the TV. Now they can turn on the TV set while the internal circuits are exposed. Always be very careful around high voltages because you conduct electricity.

Voltages are also the "signals" that contain the information in electronic systems. Very small voltages (1 to 100 μV) are provided by the antenna of a radio receiver. In contrast, signals in digital systems are relatively large voltage swings between 0 and +5 V.

Cells and Batteries

Cells and batteries (a collection of two or more cells and therefore a "battery" of cells) provide the voltage sources for portable electronic equipment. Batteries can be classified as being either made up of "primary cells," which provide electrical energy from an internal chemical reaction, or "secondary cells," which are storage elements for electrical energy. Storage batteries use secondary cells, which have to be periodically recharged (and also initially charged) to provide a source of electrical energy. Examples of storage batteries are the battery in your car (it is recharged by the alternator on the engine), and the nickel-cadmium (Ni-Cad) batteries that are being used in all types of portable equipment, even power tools. The terminal voltages of these primary cells is a function of the materials of the cell. The battery used in cars is based on a lead-acid (sulphuric acid) cell that has a terminal voltage of 2.1 V. Ni-Cad cells have a terminal voltage of 1.44 V.

In contrast to this are the primary cells, like the "flashlight battery" (which actually is a *flashlight cell*), in which an internal chemical reaction provides the electrical energy. Once the internal chemicals become depleted, the battery or cell is discarded. Again, the materials used to make the cell determine the terminal voltage. Standard flashlight cells (zinc-carbon cells) have a terminal voltage of 1.5 V. Mercury cells have a terminal voltage of 1.36 V that is fairly constant throughout the life of the cell. Silver-oxide cells are very popular for the small, single cells that are used in wristwatches and have a terminal voltage of 1.5 V.

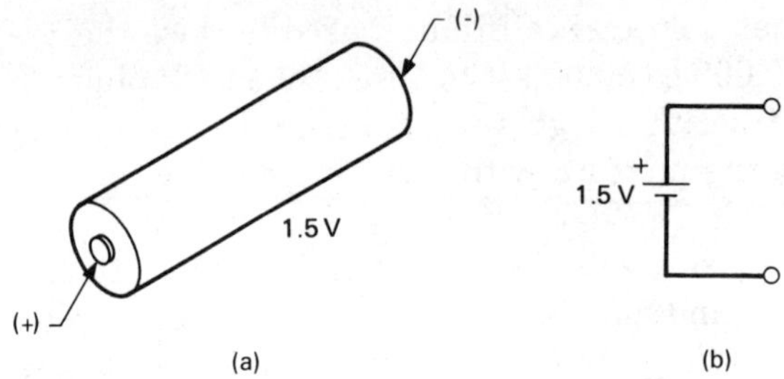

Figure 2.1 An electrical cell. (*a*) A single "flashlight cell" (these are *not* flashlight *batteries*) and (*b*) the symbol for a single cell.

The life of these primary cells depends on the way they are used. For example, a 1.5-V AA penlight cell can supply 30 mA for 14 h in continuous service or 20 mA for 33 h if used only 4 h per day. Similarly, a 1.5-V D cell (the standard flashlight cell) can supply 160 mA for 9 h in continuous service or 130 mA for 21 h if used only 4 h per day.

A single flashlight cell is shown in Figure 2.1. Also on this figure is the symbol for a single cell. A lot of special symbols are used in electronics instead of drawing an actual picture of what each electronic component really looks like.

Figure 2.2 shows that the two cells in a flashlight have to be prop-

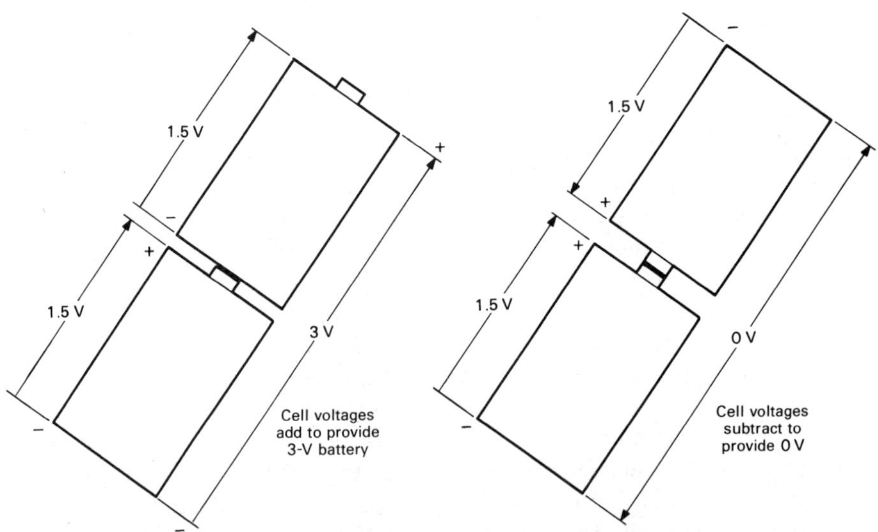

Figure 2.2 Multiple cells have to be properly oriented. (*a*) Stacking cells for a two-cell 3-V flashlight battery and (*b*) improperly stacking cells.

erly installed so that the individual voltages of each cell will add up. In Figure 2.2*a*, the button in the center on one end of the cell (the positive terminal) typically contacts with the backside of the other cell (the negative terminal). In this way the 1.5 V that is provided by each cell will add to provide 3 V for the flashlight.

In a flashlight, the batteries are the "pump" that maintains the voltage, and the current flow through the flashlight bulb generates the light that is given off by the flashlight. When you leave the flashlight on for a long interval of time, the batteries become weak because they no longer can "pump electrons up" to a potential difference of 1.5 V for each cell; therefore, the voltage across the batteries drops and this causes the current to drop and the light is no longer brilliant but rather turns yellow until finally the flashlight is not much good. Therefore, the batteries have to be replaced because they are no longer able to "pump up the electrons" to a voltage difference of 1.5 V. This is caused by the depleted chemical activity within the cell of the battery.

If the cells are placed in a flashlight in the way shown in Figure 2.2*b*, the two center buttons are in contact. The voltages of each cell therefore do not add up; they subtract to provide 0 V. If a flashlight is to be stored, put the cells in this way. Even if the switch is accidently turned ON, the cells won't be drained.

Batteries have a positive (+) side, and a negative (−) side, and it is very important to install a battery correctly to make sure that the voltage that is applied to electronic circuits is not reversed.

Some common batteries are shown in Figure 2.3. Figure 2.3*a* shows a 12-V automotive battery, a 6-V lantern battery, and a 9-V transistor-radio battery. Internally, these 6- and 9-V batteries are collections of individual cells in which each cell provides 1.5 V. The number of cells that are connected in series determines the voltage of the battery, and the physical size of the individual cells determines the amount of current which can be taken from the battery. Batteries for lights, such as flashlights and lanterns, are relatively large because of the large amount of current that is required. In contrast, a 9-V transistor-radio battery uses six relatively small-sized cells, because the current drain is relatively small.

The symbol for a battery is shown in Figure 2.3*b*. It is seen as simply a stack of cells. The actual number of cells that is used in the battery is not shown. The voltage of the battery is simply written next to the symbol.

Figure 2.4 shows the analogy between a hydraulic circuit and a flashlight, an electric circuit. A hydraulic circuit, Figure 2.4*a*, makes use of water in an elevated storage tank. The height of this water creates pressure, and it is this water pressure that makes the water flow.

Typically a pipe or a hose is used to carry the water to where it will be used. A relatively large diameter hose will prevent losing much of

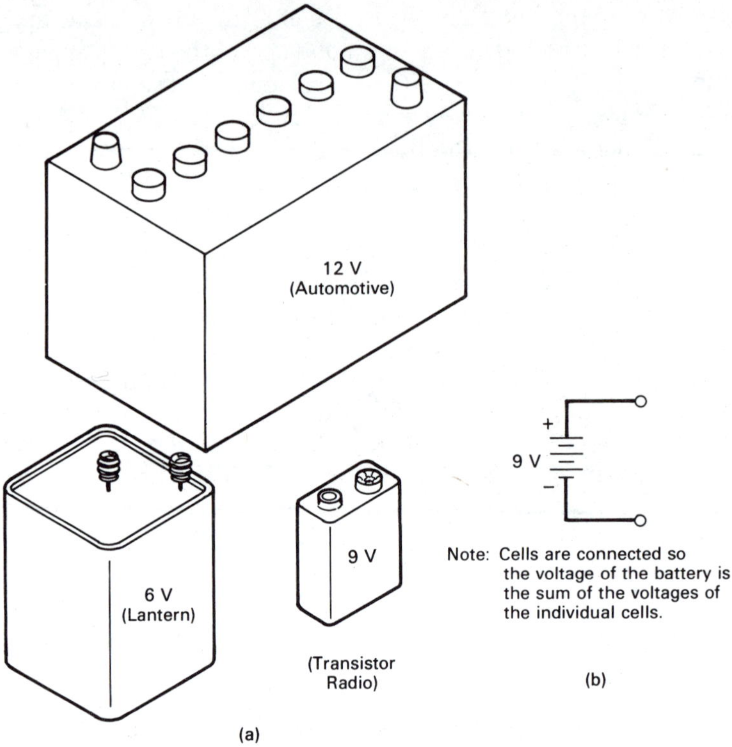

Figure 2.3 Batteries are groupings of more than one cell. (*a*) Some common batteries, (*b*) the symbol for a battery (a number of cells).

the water pressure. A spray nozzle is shown on the end of the hose. An ON/OFF valve is used to turn the water flow on or off; this is the ON/OFF switch for the water system.

The nozzle is the main restriction, or "resistance," to the flow of water. The pressure that results from the height of the water tank is pushing the water through this nozzle. The thing that is moving in this hydraulic circuit is water.

With this hydraulic circuit in mind, let's now consider an analogous electric circuit: a flashlight. The two cells in the flashlight example shown in Figure 2.4*b* add up to provide 3 V. This voltage is the "pressure" in this electric circuit. There is also an ON/OFF switch included in this circuit diagram. The electrical symbol for the lamp used in the flashlight is shown.

When the ON/OFF switch is closed, the thing that will flow through this circuit is "current," and the current flowing through the resistance of the light bulb creates the light. A complete circuit connection is needed, going from the positive terminal of the upper cell, through the

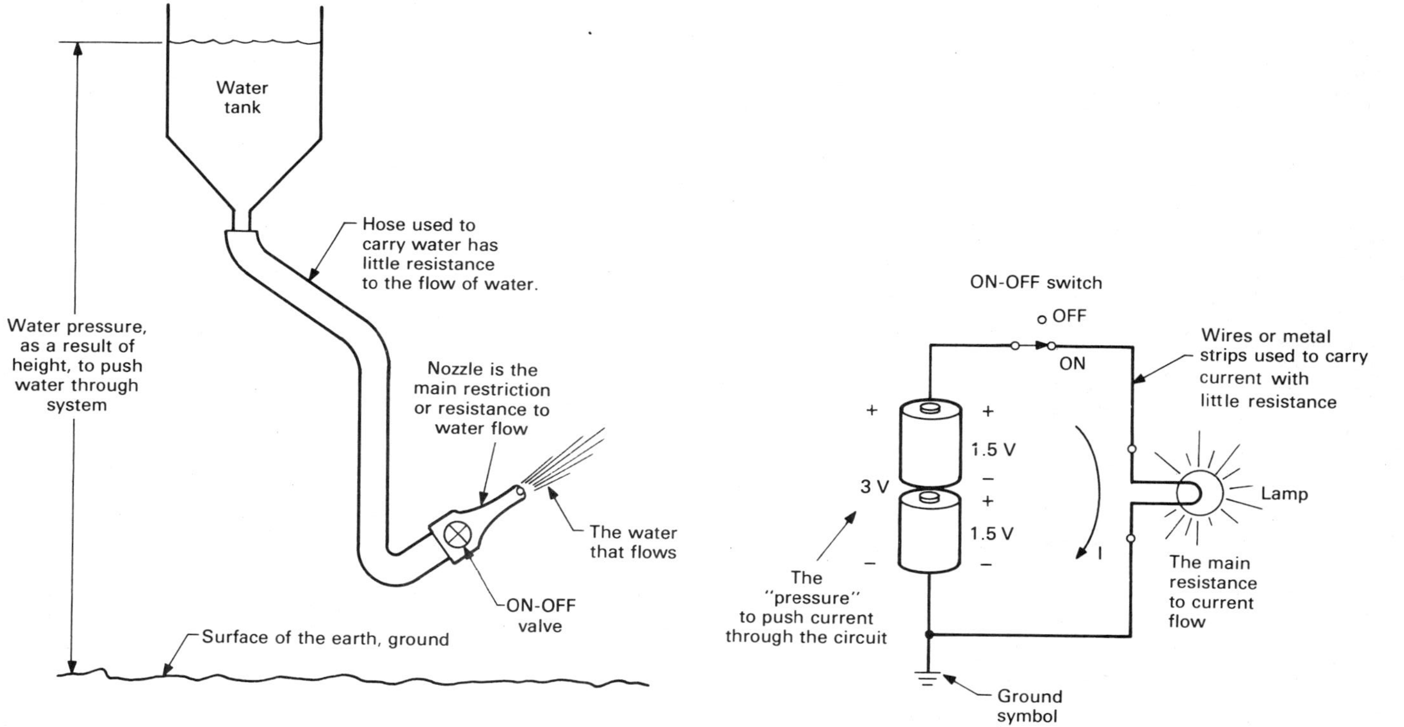

Figure 2.4 The hydraulic analogy of a flashlight. (*a*) A water circuit and (*b*) the electrical circuit of a flashlight.

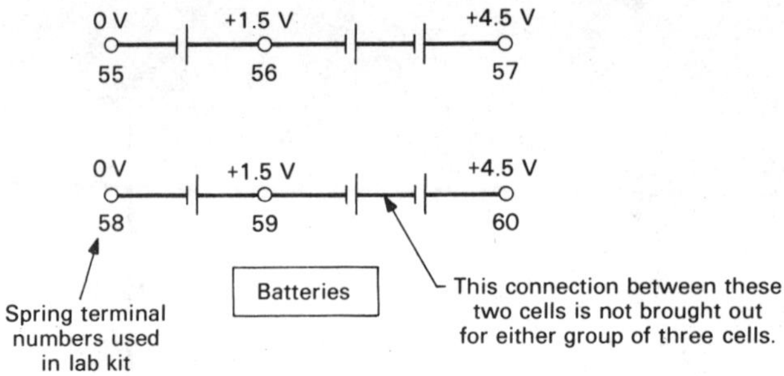

Figure 2.5 Connections to the six AA cells of the lab kit.

switch, through the light bulb, and then back down to the negative side of the lower cell. If this electric circuit is interrupted (by opening the switch), current can't flow. To allow current to flow, a continuous loop is needed, a closed circuit.

In the upper right-hand corner of the lab kit are six AA cells. Contacts to these cells are available through six terminals in the area labeled "batteries," as shown in Figure 2.5. This group of six cells provides the voltage sources for the experiments that use the lab kit.

Ways to Pump Up Electrons: The Generation of Electricity

To generate electricity, ways must be found to take electrons from one place and move them to another place ("pump them up") so a voltage will be established that causes them to return. We'll now take a look at a number of ways to generate electrical energy.

Stealing electrons: triboelectricity

"*Triboelectric voltage generation*" takes place when dissimilar materials (one or even both can be insulators) that were once in contact are separated. In this separation, electric charge is transferred. The surprising thing is that two plastics (both of which are good insulators) that are rubbed together and then separated can generate triboelectric voltages.

The simplest example of triboelectric voltage generation occurs when a cat's fur is brushed on a dry day. The cat's fur will stand up and will also respond to the position of the brush. Through a triboelectric generation phenomenon, electrons have been removed from the cat's fur and have been transferred to the comb. A spark discharge can occur as the brush approaches the cat's fur again.

Triboelectric voltage generation has caused many problems for semiconductors because integrated circuits are shipped within plastic carriers. As the integrated circuits slide in and out of these plastic tubes they can generate enough triboelectric voltage to permanently damage the integrated circuits. All of these tubes now have carbon mixed in with the plastic so that they are conductive plastics and will therefore not generate triboelectric voltages.

The new flintless, batteryless cigarette lighters: piezoelectricity

Another way to generate a voltage is to apply a mechanical force to distort a crystal. Voltages that are generated within the crystal structure are trying to restore the crystal to its equilibrium position. This electrical generation phenomenon is "piezoelectricity" and has been used in crystal microphones, earpieces (as in the lab kit), phonograph cartridges, and crystal frequency references.

A phonograph needle, when attached to a piezoelectric crystal, will mechanically stress the crystal as the needle follows the record grooves. This can generate a few volts of electricity.

A more modern application of piezoelectricity has provided a new way to light a cigarette. No longer are flints needed. By mechanically distorting a crystal, a voltage is generated, and the resulting electric spark ignites the flame of the cigarette lighter.

Niagara Falls and rotating machines

Whenever water is moving, or is dammed up and the escape can be controlled, the motion of this water can be made to turn a wheel. At Niagara Falls the water flow over the falls causes large turbines to rotate. Alternators (ac generators) then convert this rotary motion into electrical energy.

These hydroelectric sources of electricity are very important for a lot of America's electrical energy requirements. Hydroelectricity has the benefit of not polluting.

Solar cells: The key to the batteryless calculators

Solar cells on the roofs of homes convert sunlight into electrical energy. Another very common example of the solar cell is the light-powered calculator. No longer are batteries necessary to power a hand-held calculator. If the light is bright enough to allow you to read, there is enough light to operate these light-powered calculators. Both four-function calculators and scientific calculators can be powered from ambient light. This is quite a breakthrough.

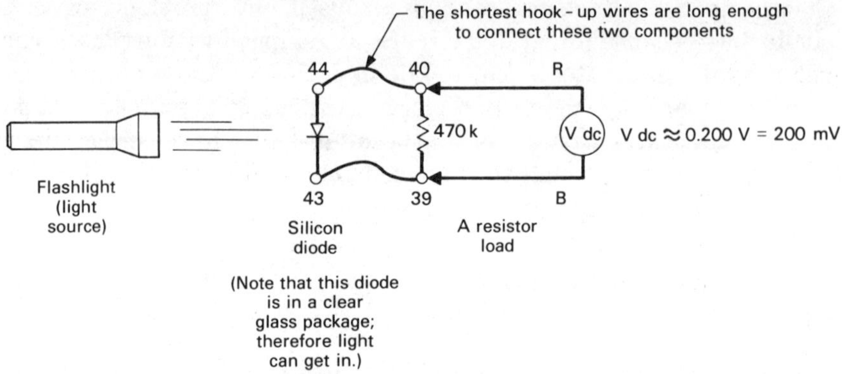

Figure 2.6 Converting light into voltage: the solar cell.

The silicon diodes in the lab kit can be used as examples of solar cells. These diodes are packaged in clear glass, and if a flashlight is shone onto these diodes, they will generate a voltage. This is the "photovoltaic effect" that exists with semiconductor diodes.

A resistor can be connected across one of these silicon diodes, as shown in Figure 2.6. The voltmeter measures the voltage across both the diode and the resistor (the use of the voltmeter is discussed in the next section). If light does not shine on the diode, the voltage reading will go to zero. With light, a voltage of approximately 200 mV (0.200 V) can be obtained.

If both diodes were connected in series (and a light was shone on both of these diodes) the 200 mV from each of the solar cells would add up. This is the way higher voltages are obtained from arrays of solar cells.

Electrical power generation for the future is expected to make increasing use of the sun, water, wind, cogeneration (use fuel more efficiently to generate electricity and also produce steam or heat for other uses), and geothermal (using the inner heat of the earth) steam.

Measuring DC Voltages

While considering voltage, the "digital multimeter" will be introduced, because this can be used to measure the value of a dc voltage. A particular Micronta model (Radio Shack Cat. No. 22-188) is used as an example, but many other voltmeters or multimeters exist and can also be used (the arrangement of the front panel will probably be different, and therefore the owner's manual should be consulted to determine the proper setup to obtain the same measurement function).

The way to set up the digital multimeter to measure dc volts is shown in Figure 2.7. Notice that in the middle of the multimeter is a rotary

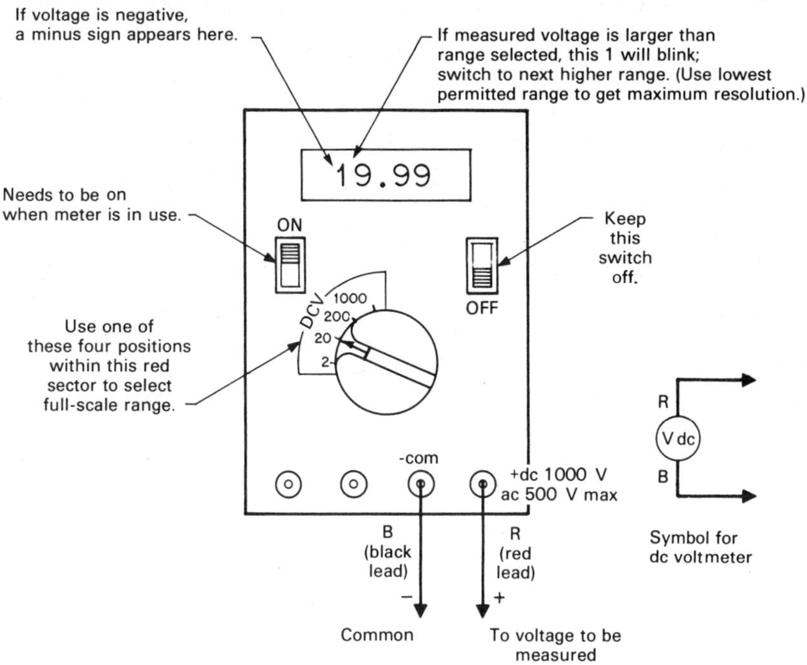

Figure 2.7 Setting up the digital multimeter to read dc volts (V dc).

switch, and there are five different types of measurement functions that are provided by this meter: DC V (the ability to read dc voltages), AC V (the ability to read ac voltages), "diode" CHECK (the ability to test semiconductor diodes), KΩ (the ability to measure the value of resistance of resistors and also the continuity of wires), and DC mA (the ability to measure dc current flow). To measure dc voltages, switch the center selector switch over to the left-most position that is labeled "DC V." The meter is "autoranging" so it will automatically determine the proper range (out of a possibility of five ranges) to use for the voltage that is being measured. For small voltages the range will be 200 mV full scale and "mV" will be displayed in the upper right-hand corner of the display. The four other full-scale ranges that can be automatically selected by the meter are 2 V, 20 V, 200 V, and 2000 V. For these ranges, the decimal point will be automatically located, the "mV" will not appear in the display, and the voltages are directly read in volts. We will be using only the first three ranges, 200 mV, 2 V, and 20 V, in the experiments that are shown in this book.

To attach the meter probes to the meter, notice that the black meter lead always plugs into the center jack, which is called "−COM" (minus Common). The red lead will go to the right of the black lead to measure both dc and ac volts and, as shown later, it will go to the left of the black lead for reading current (dc milliamps) or resistance.

To set up the multimeter to read dc volts, point the arrow of the selector to the left DC V position. Before a reading can be obtained, the power switch on the upper-left edge of the meter must be turned on. The multimeter is now set up to read dc volts and will automatically provide an easy-to-read display of the voltage value.

A negative sign may appear to the left of the digits in the display. This negative sign simply means that the polarity of the dc voltage being measured is different from what you thought. Typically, the black lead is connected to the negative side or ground side of the measured voltage and the red lead is connected to the positive side. The reading will then be a displayed number; there will be no algebraic sign displayed. This implies that the polarity of the measured voltage is the way you thought it was. In some circuits, both positive and negative voltages exist, and this polarity indication is then a useful feature because the leads don't have to be interchanged between readings.

When using this meter to measure voltages in the lab kit, notice that the tips of the meter probes nicely fit inside of the contact springs of the lab kit. The probe will just lay sideways making an adequate contact, as shown in Figure 2.8. There is a large mechanical lever arm here, so don't pull sideways on the meter probe because the contact springs of the lab kit can be damaged.

Two different ways to connect the probes of the meter to read the voltage that exists between terminals 58 and 59 of the battery section are shown in Figure 2.9. The meter will indicate the voltage of a single cell, a value of about 1.5 V.

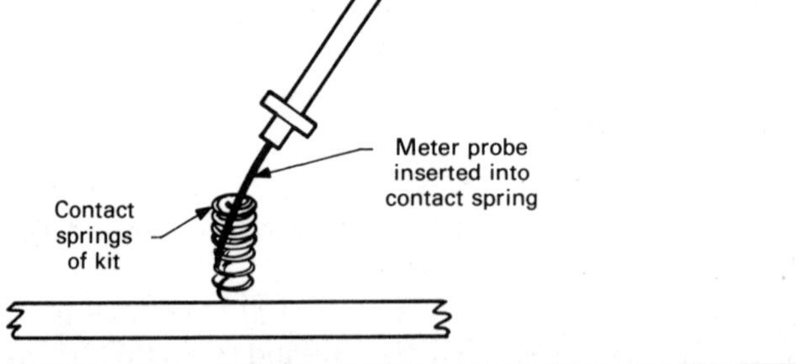

Figure 2.8 Meter probes easily fit into the contact springs of the lab kit.

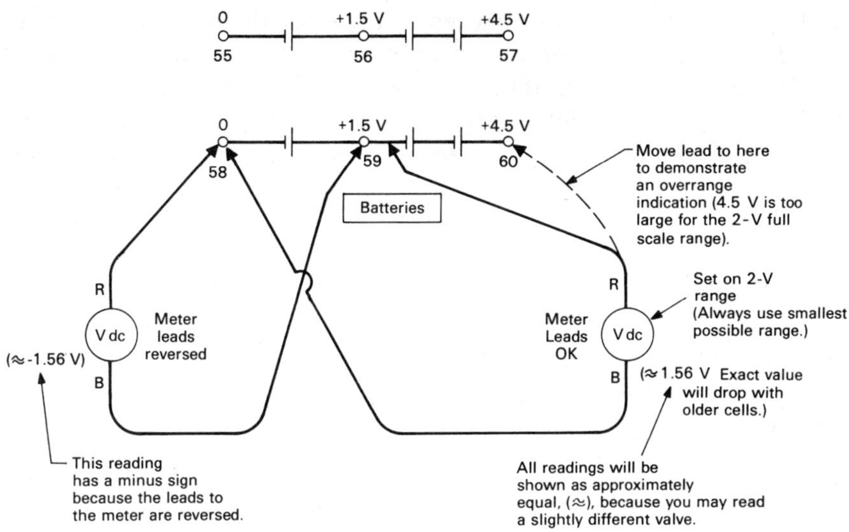

Figure 2.9 Measuring the cell voltages in the battery section of the lab kit.

Whenever a voltage reading is mentioned, it is hard to predict the exact value that will be read. Therefore, the approximately equal sign (-) is used. The actual reading should be very close to the value shown. One of the reasons for a difference in the readings is that the voltage of the individual cells that are used in the battery section tends to fall as they are used. You have experienced this with a flashlight. If the batteries are too old, they provide less voltage, and therefore the brightness of the flashlight is reduced (the light also turns yellowish instead of being a bright white). Old batteries in the lab kit will also give lower voltage readings. In fact, this voltmeter can be used to determine if a battery or cell is good or not. It isn't necessary to put a cell into a flashlight or to put a battery into an electronic gadget to determine its condition; simply read the terminal voltage. A low-voltage reading (less than 1.2 V per cell) indicates that the cell or battery should be discarded or recharged.

On the right side of Figure 2.9, notice that the red lead is connected to terminal 59, and the black lead is connected to terminal 58. The meter will indicate a positive voltage very close to 1.56 V with fresh cells. The meter can indicate as low as 0.7 or 0.8 V if the cell is beyond the end of its life.

Now simply reverse the leads, as shown on the left side of Figure 2.9. Connect the red lead to terminal 58 and the black lead to terminal 59. This connects the leads in the reverse order so the meter indication will now be −1.56 V. This shows what the minus sign looks like. If this minus indication is seen, it means that the polarity of the voltage being measured is negative. (Also note that if the red lead is contacted to the negative side of a measured voltage and the black lead is contacted to the positive side, this negative indication will also appear.)

The cells that are available in the lab kit can be used in many ways. In Figure 2.10 a short jumper wire (use the shortest wire available) is shown going from terminal 55 to terminal 60 (always leave this jumper wire in place). If the lower left-hand terminal (58) is used as the common point (or the "ground point"; there will be more discussions about ground later) and the black meter lead is contacted to this point, the red lead can be touched to any of the available terminals in the battery area to measure all of the positive voltages that are available. This ground connection allows all of the cells to be added up in series to provide voltages larger than 4.5 V.

If the red meter lead is contacted to terminals 55 and 60, the meter will indicate 4.5 V. This is the first three cells added together. If the red lead is moved to terminal 56, another 1.5 V are added to provide about 6 V. Touching the red lead to terminal 57 adds up the two 4.5-V sections to provide 9 V. Touching the red lead to terminal 59 will provide an indication of 1.5 V, the voltage of one cell. Write down the actual voltages measured with the cells that are in your kit.

Positive and negative voltages can be simultaneously provided if the jumper wire (the one that ties terminal 55 to 60) is considered to be

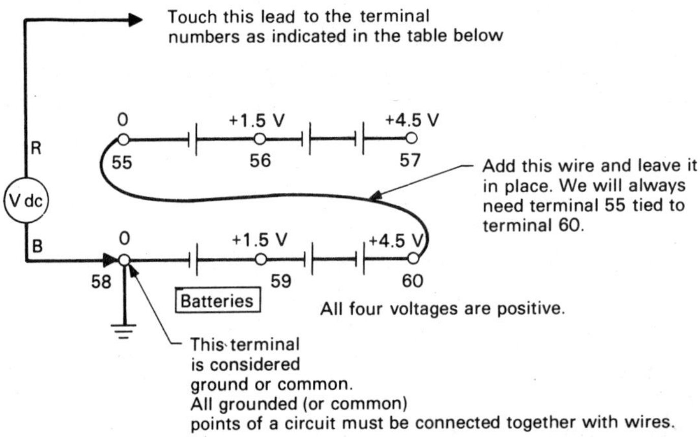

Terminal number	Nominal voltage (V)
58	0
59	1.5
55 & 60	4.5
56	6.0
57	9.0

Figure 2.10 Obtaining multiple positive voltages in the lab kit.

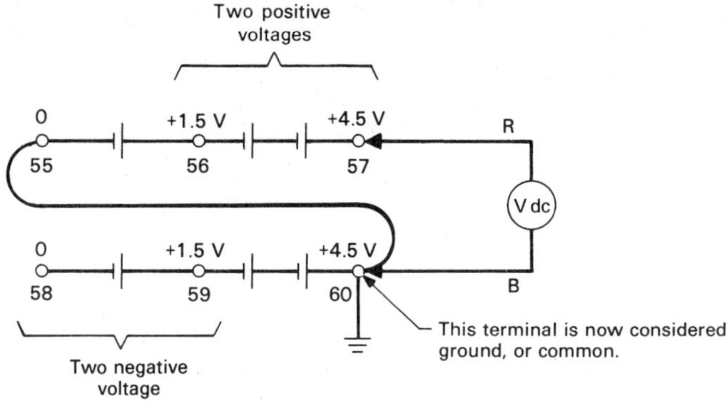

Pin number	Nominal voltage
55 & 60	0
56	+1.5
57	+4.5
58	-4.5
59	-3.0

Figure 2.11 Obtaining multiple positive and negative voltages in the lab kit.

the common, or ground, point. Connect the black lead of the voltmeter to this terminal. This provides two positive voltages (on terminals 56 and 57) and two negative voltages (on terminals 58 and 59), as shown in Figure 2.11. The nominal voltages that should be read are also shown. Notice that the black lead is tied to the ground point (terminals 55 and 60); therefore, when negative voltages are measured, there will be a negative sign in the display on the meter. This ground connection will be used for the electronic experiments that use both +4.5 and −4.5 V.

Determining the Voltage Needed to Operate a Relay

Relays are commonly used in both electric and electronic systems. A relay is an electrically actuated switch. The benefit of using a relay is that the large-diameter wires required for a high current load do not have to be run over large distances to get to where the switching control takes place. Also, the high current load can be remotely turned on and off with a low-power control circuit.

When a relay closes or opens, it can perform a switching function on a large current flow. Relays therefore have "power gain"; a low-power circuit is used to control a high-power circuit. For example, every car uses a relay to operate the starter. The starter is an electric motor that is used to turn the engine over in order to get the engine running. When the ignition switch engages the starter, a starter solenoid is energized. This solenoid is a relay that switches the large currents that are needed by the starter motor. This eliminates the cost of long large-diameter battery cables that would otherwise have to be routed into the drivers' area, and it also eliminates a large, high-current manually operated switch that otherwise would also be needed.

There is a relay in the lab kit that is the large component that is located in the upper central area. Figure 2.12 shows connections to the control coil of this relay via terminals 61 and 62. Wires are shown connecting from these terminals over to the batteries. Notice that terminal 62 is connected to terminal 58 with one wire, and another wire is used to contact the other terminal (61) of the relay to different voltages to determine how much voltage is needed to operate this relay.

The relay will produce an audible click when sufficient voltage is applied to the control coil. This control coil sets up a magnetic attraction that moves the armature and thereby switches the mechanically attached relay contacts (terminals 63 and 64 open, and terminals 64 and 65 close).

First, take the wire from terminal 61 and touch it to terminal 59. This applies 1.5 V to the control coil and, as indicated in the table of Figure 2.12, the relay will not operate. Now, move that wire from terminal 59 to terminal 60 (applying 4.5 V to the relay coil). The relay still won't operate. But, when the wire is moved up to terminal 56 (applying 6 V) a clicking noise will be heard, indicating that the relay has operated. Another click will be heard when this wire is removed. This clicking is the noise made by the moving armature as it opens and closes the relay contacts. If the wire is contacted to terminal 57 (applying 9 V) the relay will also operate.

In the symbolic drawing for this relay, the contact between terminals 63 and 64 is shown closed when the control coil is not energized. When 9 V is applied to the control coil of the relay, the center element of the relay (called the "armature") will move down. This movement will close terminals 64 and 65 and open up terminals 63 and 64.

A typical use for a relay is shown in Figure 2.13. Notice that many hook-up wires are needed here. The Key switch (available in the lab kit) is also used. Closing this switch will actuate the relay. The LED lamp (terminals 169 and 170) is also used and is controlled by the normally open set of contacts of the relay (terminals 64 and 65).

Notice that the 3 V for the lamp comes from terminals 56 and 57 of the battery section. Terminal 169 of the lamp goes to a 100-Ω resistor and then to terminal 65 of the relay contacts. Terminal 64 of the relay contacts ties to terminal 56 of the battery section. Terminals 56 and 57 are contacting two cells in series to provide 3 V for the lamp.

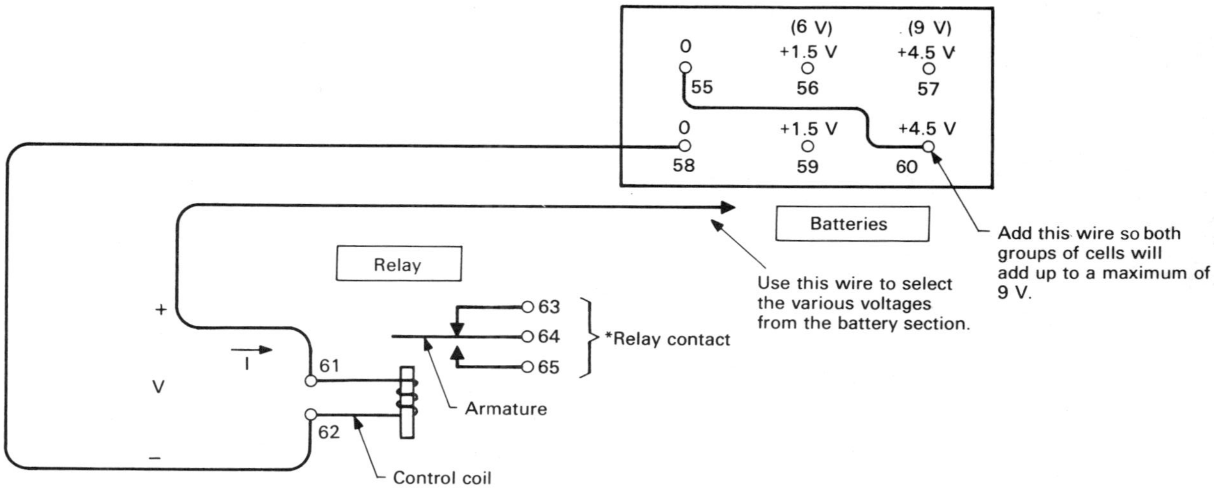

*Terminals 63 and 64 are normally closed and terminals 64 and 65 are normally open. This interchanges when 9 V is applied to the control coil.

Pin #	V	Are contacts open?
59	1.5 V	No
60	4.5 V	No
56	6.0 V	Yes
57	9.0 V	Yes

Figure 2.12 Determining the voltage needed to operate the relay in the lab kit.

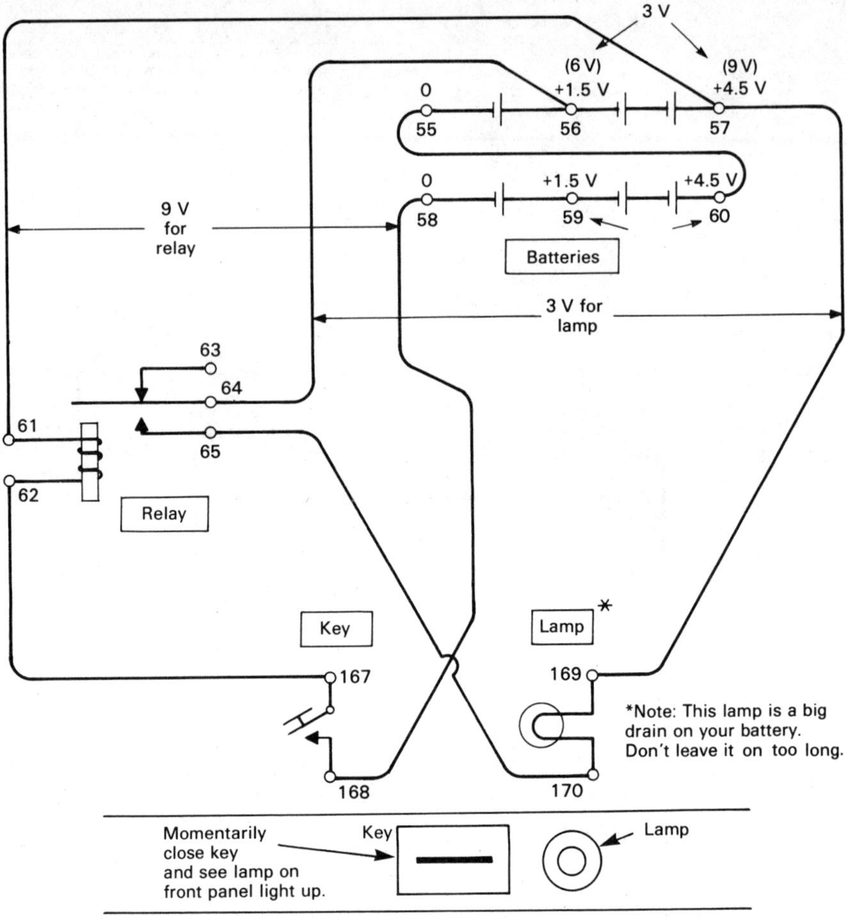

Figure 2.13 Using the relay to operate a lamp.

The control coil of the relay is powered across all of the cells in series to provide 9 V. Terminal 58 of the battery section goes to terminal 168 of the Key switch. Terminal 167 of the Key switch then goes over to the relay coil (terminal 62). The other relay coil terminal (61) goes up to terminal 57 of the battery section.

When the red Key switch is pushed down (closed), voltage is applied to the control coil of the relay. This will cause the resulting armature movement of the relay to connect terminal 64 to terminal 65, which completes the electric circuit to light the lamp. Depress the Key switch to verify that the lamp lights when this switch is closed.

Now that voltages have been discussed, we can move on and take a look at some "analog," or "linear," systems.

Chapter

3

What Are Analog Systems?

You will hear about analog systems or linear systems and also digital systems or computer systems. This is the big separation within electronics, analog versus digital. The analog, or linear, systems use voltages that "wiggle around." Digital systems make use of a binary system that requires only two voltage levels, either 0 or 5 V.

Analog systems make use of voltages to represent various physical measurements. A voltage magnitude can represent things like temperature, humidity, pressure, the amount of light, or even the position of an object. Voltages represent measurements in an analogous form. For example, a high voltage would represent a high temperature. Another voltage in an analog system could represent a particular pressure, such as fuel-line pressure in an automobile. A wide range of pressures would then be represented by a wide range in the magnitude of this analog voltage. Other voltages could represent position, velocity, or even acceleration. That's the reason the word "analog" is used for this type of an electronic system—voltage analogs of physical measurements are being processed by linear (or analog, a synonym) electronic circuits to provide an analog control or measurement function.

Some Traditional Analog Electronic Systems

In your normal range of experience you have come across many analog systems. A simple way to classify electronic systems is that analog systems consist of all of the systems that are not specifically a digital or a computer-type system. Today, there is an increasing tendency to combine both analog and digital circuits within the same electronic system.

The earliest electronic systems were all analog systems, and as time went on, more and more of these traditional analog systems were converted to digital. That's why you hear of a takeover by digital electronics and computers today.

Many consumer products make use of linear systems; they include radios: AM, FM, and citizen's band transmitters/receivers ("transceivers"). Today, digital techniques are being added, especially in the "frequency synthesizers," that allow exact tuning to a desired station (for both transmitting and receiving). Television sets have always been a combination of both analog and digital circuits, and digital techniques are being expanded to provide many of the TV circuit functions that have, in the past, been handled by analog circuits.

Digital techniques have been applied to the traditionally linear high-fidelity sound reproduction systems with the compact disk systems, and digital techniques are even appearing in one of the first high-volume analog systems: the telephone system. Now voice signals are converted into "digital words" before they are transmitted over the telephone network. These digital signals are then converted back to an analog signal just prior to entering the handset.

Telephones exist all over the world and provide a convenient way to communicate from any point to any other one. So the obvious thing is to find ways to make use of this vast network of telephone lines to increase its capacity for all forms of communication.

The requirement for communications continues to increase daily. In addition to more and more people wanting telephone service, digital computers and other services are also communicating over the existing wires of the telephone network. This ever-increasing demand for service has required a different technique for the transmission of speech. For example, to double the service capability of the telephone network would require doubling the existing telephone cables, and twice the number of central offices would also be needed. This is too costly, so the approach that is taken is to use relatively complex integrated circuits to allow a large number of simultaneous telephone conversations (or other types of communication) to be placed on an existing pair of wires. In this way, the service capability is increased by installing complex electronics and using the existing buildings and the network of telephone lines. It's the relatively low cost and the sophisticated performance that integrated circuits provide that is allowing this sophisticated electronic solution to the demands for increased telephone service.

There are still a number of analog systems available, but there is a general tendency to convert from analog to digital in many modern systems.

All electronic systems, whether they are digital or analog, are assembled from basic circuits, and these circuits are, in turn, made up of electronic components. Figure 3.1 shows the symbols for a number of these components. (Electronic components will be described in more detail in Chapter 4; this is just an indication of the components that are available and the symbols that are used to represent them.) Complete linear circuit functions are provided by transistors and special linear integrated circuits (ICs) and these are used in combination with the components that were shown in Figure 3.1 to make up linear circuits and linear systems.

Electronic systems make use of schematic diagrams that are a symbolic representation of the circuit details. Schematic diagrams make use of electronic symbols to indicate the details of how the circuits are made, the values of the components, and also how the individual circuits are interconnected to make the complete electronic system. These diagrams are necessary in order to understand what is going on (for example, when repairs are needed).

To construct an electronic system or a circuit, the leads of all of the components are soldered to a supporting board, a printed circuit (PC) board. The benefit of a PC board is that the interconnect wiring does not have to be done by the high-cost technique of soldering many wires

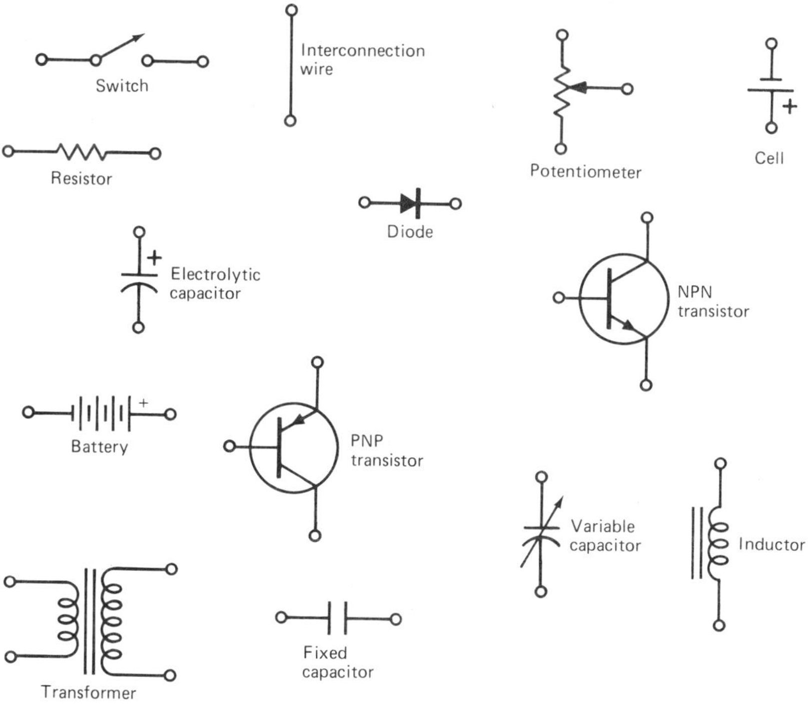

Figure 3.1 Electronic systems are assembled from these components.

in place by hand. This circuit interconnection is accomplished with copper traces that are available on one or both sides of the PC board.

On one end of a PC board is a special connector so that the board can be slipped into a multipin socket that will allow electrical contact to the copper traces on the board.

These boards are improperly called "printed" circuit boards; this copper foil is not put in place using a *printing* operation. The copper interconnect traces have resulted from an etching step that dissolved away most of the copper on a completely copper-cladded board. Only a particular copper interconnect pattern was allowed to remain. So, a more proper name is an "etched" circuit board.

Analog systems generally require a lot of external components to be used around the integrated circuits. Digital ICs, on the other hand, can simply be interconnected directly and not as many external resistors, capacitors, and other electronic components are needed. When a PC board contains a lot of resistors, capacitors, transformers, and diodes, it is part of a linear system. If, on the other hand, there are only rows and rows of integrated circuits all connected together, with only a few capacitors here and there, it is a digital PC board.

Obtaining Voltages that Represent Physical Things

Various gauges are available that allow measurements to be made of physical parameters. For example, there are special pressure gauges that measure the air pressure that exists in the tires in a bicycle or an automobile. Similarly, there are temperature gauges to measure temperature. These types of gauges require a human to take the readings. Electronic systems make use of many types of transducers to convert physical measurements into analog voltages because voltages are the inputs to analog electronic control systems. For example, an automatic controller for room temperature monitors room temperature, as measured by a transducer, and the temperature is converted into a voltage analog. This electrical signal is then used within an electronic control system to determine whether or not the heater should be turned ON.

Microphones are also transducers; they convert the variable pressures that are generated in human speech into a "wiggling" voltage. Loudspeakers are transducers; they convert electrical voltages to sound waves in the air. Other transducers exist that measure pressure and even displacements (small movements of an object from a reference position).

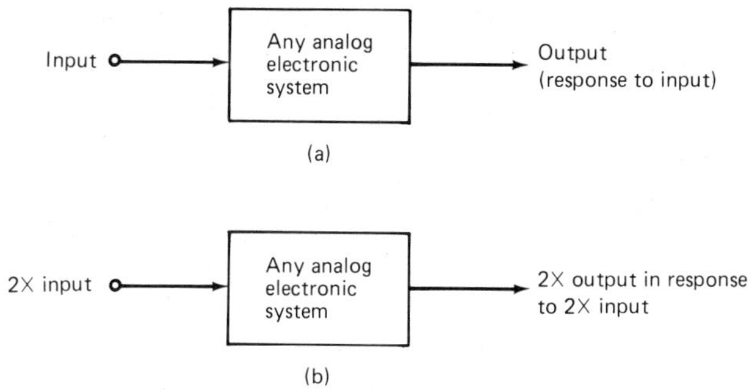

Figure 3.2 The meaning of linearity. (*a*) A basic representation and (b) the definition of linearity.

The Meaning of Linearity

In analog systems, there is a big interest in obtaining what is called a "linear system." The basic definition of a linear system is indicated in Figure 3.2, in which an analog electronic system is shown that has an input and an output. This particular analog system is said to be linear if, when we double the input (increase the input by a factor of two), the output also increases by a factor of 2. The output would then be responding in a linear way to the input.

Linearity is required in a hi-fi audio system. The input in this case is coming from a tape, a phonograph record, or an FM or AM tuner. A linear system ensures that the amplifiers in the hi-fi don't cause distortion. It will faithfully provide an output in response to an input, and it will not introduce any new signals or other distortions. The goal is to have the amplifier only increase the signal level and not distort or change the input signal in any other ways.

Chapter

4

Circuit Components Operating with Direct Current

In the discussions about circuit components, we will limit our concern to operating with direct current. Direct current (dc) is used to imply voltages that are reasonably constant in magnitude and do not change signs or currents that are reasonably constant and do not change direction. Alternating currents (ac) are currents or voltages that are constantly changing with time. For example, in homes the wall electrical outlets provide 110 v ac. Batteries, on the other hand, provide dc voltages.

There are many electronic components and items of hardware that will not be discussed in detail in this book, such as lamps, switches, connectors, plugs, sockets, jacks, heat sinks, fans, and PC board materials. Many people devote their careers to the understanding, design, and manufacture of electronic components. Another group of people (component engineers) work as specialists within engineering organizations to keep up to date with the latest components, and these component engineers consult with design engineers on component applications.

Resistors: To Resist Current Flow

A very common electronic component is the resistor. Figure 4.1 shows the electronic symbols for various types of resistors. A "fixed resistor" is shown in Figure 4.1*a*. This is a component that is designed to have a specific amount of resistance. In the center of this drawing is a variable resistor, a "rheostat." A rheostat allows the resistance

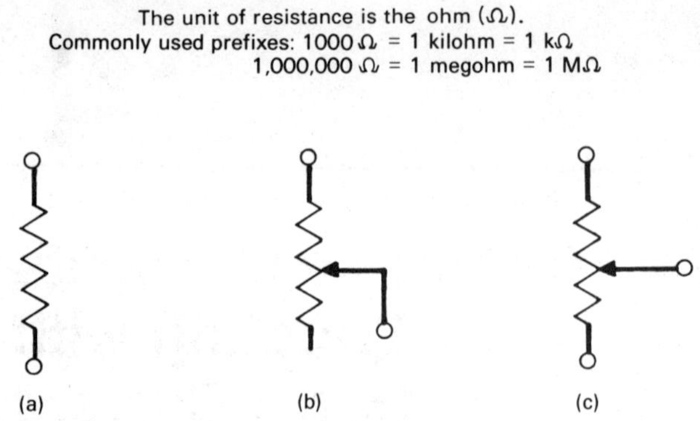

Figure 4.1 Resistor symbols. (*a*) A fixed resistor, (*b*) a variable resistor (a rheostat), and (*c*) a variable tap on a fixed resistor (a potentiometer).

value to be varied, either by turning a screw or a knob or by loosening a clamp and sliding a wiper physically up and down the resistor element.

The last thing shown in this figure is a fixed-resistor element with a sliding wiper, a "potentiometer." This wiper provides a voltage that lies between the voltages that are placed across the main resistor element. The wiper can therefore sample the voltage from one end of the resistor element to the other end. Potentiometers are used as volume controls and tone controls in televisions and radios. To increase the audio level, the wiper is moved toward the higher audio signal voltage that exists at one end of the resistor. To reduce the audio level, the wiper is moved in the opposite direction, to the end that has no signal, the grounded end.

The unit of resistance is the ohm (Ω). Resistors commonly make use of metric prefixes, such as kilo (kilo ohms, or 1000 Ω, sometimes referred to simply as k and also kΩ) and mega (mega ohms, MΩ, 1 million ohms, referred to as meg and also megohm).

Figure 4.2 shows some of the various types of resistors that exist. Even a much wider range of resistors than are shown here are available. Resistors come in many different physical sizes and are also made out of various resistive materials.

Resistors are also very common within the appliances that are in homes. Examples are electric kitchen ranges, electric stoves, electric ovens, bathroom heaters, electric hot water heaters, hair dryers, hair curling irons, electric coffee pots, heat lamps, waffle irons, and electric clothes dryers. All of these, even incandescent light bulbs, are basically resistors. These resistors are used to convert electrical energy into thermal energy, heat.

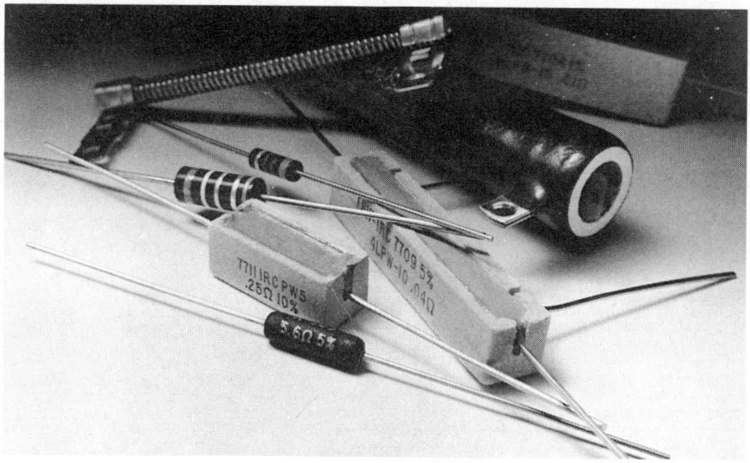

Figure 4.2 Resistors look like this. *(Courtesy of TRW Resistor Products Division.)*

The visible light that comes out of an incandescent light bulb is almost incidental as far as making use of the available electrical energy is concerned. The resistance of the filament wire inside of the glass bulb causes this wire to become extremely hot. When the filament is at about 2000°C, the filament radiates white light. Therefore, this very intensely heated resistor (the filament) serves as a lamp. The problem is that most of the electrical energy that is provided to an incandescent lamp is radiated as heat. The heat is not desired; we want only the light. Fluorescent lights are much more efficient: more of the electrical energy that is provided is converted into usable light.

The electrical line cords that carry electricity to appliances are also resistors. Figure 4.3 shows a number of the different cords that are used in a home. At the top of this illustration is the very thin, small-diameter, electrical cord that is called "lamp cord" or "zip cord." It is used for the small current flow of radios, table lamps, and floor lamps. Zip cord is one of the smallest electrical cords found in homes. Some of the heavier current drain appliances such as hair dryers, toasters, and waffle irons use larger-diameter electrical cords. More strands of copper wire are used to reduce the resistance in the electrical cords of these high-current-consuming appliances. These large currents make the line cords warm and also cause an undesired drop in the voltage that is actually supplied to the appliance.

Wires are also used to interconnect circuit components because wires have a relatively low resistance. In fact, the resistance of the interconnecting wires is usually low enough that it can be neglected; the wires are assumed to be perfect conductors.

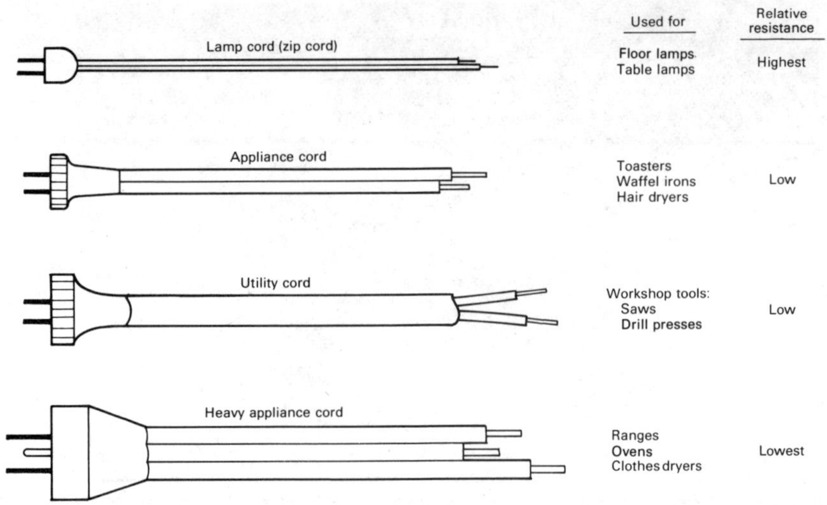

Figure 4.3 Wires have resistance too.

Resistors come in various tolerances, a way to indicate how precise the value of the resistance is. Common resistors for use in electronic circuits are supplied with a 5 percent tolerance. This means if a resistor is marked 100 Ω, it can actually measure anywhere from 95 Ω (5 percent low) to 105 Ω (5 percent high). Resistors with a 1 percent tolerance are also available. Therefore, a 100 Ω, 1 percent resistor would measure somewhere between 99 and 101 Ω. In addition, 0.1 and 0.01 percent tolerances are available but are more expensive.

Because of this tolerance, resistors with every possible numerical value of resistance are not made available. There are standard multipliers for 5 percent resistors that are shown in Figure 4.4. This is much like the scientific number system. These standard multipliers are not used to indicate the scale factor. They show what values are available for the numerical part, not the factors of 10. The standard multipliers range from 10 to 91 and they make discrete steps in a sequence such as 36, 39, 43 for 5 percent resistors. This is implying (as shown on the figure) that 10-, 130-, or 240-Ω 5 percent resistors are available but not 3.4 k or 14 or 158 Ω. Lower-cost resistors are also available with a broader tolerance allowance of 10 or even 20 percent. Ten percent resistors use a reduced number of standard multipliers; 20 percent resistors use even fewer. In contrast, the number of 1 percent standard multipliers is much larger, so there is a much broader range of values available with 1 percent tolerance resistors.

What is considered a large value of resistance and what is considered a small value of resistance? As shown in Figure 4.5, this question

Standard multiples for 5% tolerance
10
11
12
13
15
16
18
20
22
24
27
30
33
36
39
43
47
51
56
62
68
75
82
91

This list of standard multipliers can have any decimal multiplier over the commonly supplied range of 10 Ω to 10 MΩ.

Therefore 5% resistors have values as:

10 Ω	130 Ω	200 Ω
2.2 kΩ	3 kΩ	3.6 kΩ
39 kΩ	43 kΩ	51 kΩ
620 kΩ	750 kΩ	1 MΩ
1.8 MΩ		

— not 14 Ω, 158 Ω, 660 k, 3.4 k, etc.

Note: A 1-kΩ ±5% tolerance resistor = 950 Ω to 1050 Ω in actual value.

Figure 4.4 The standard resistor values for 5 percent tolerance.

will be answered differently by different people. For example, electric power people who work with big hydroelectric generators, electric power lines, and electric power distribution stations would consider something like 0.1 Ω a large resistance value. Large values of current exist so the resistance of the power lines must be kept very small to prevent energy loss as wasted heat.

An electrician who works on the wiring in homes would consider 1 Ω a large value of resistance. A linear circuit designer would say that resistors with values larger than 500 kΩ would be large (on an IC chip

Person	Resistance valves greater than
The electric power worker	0.1 Ω (one-tenth of an ohm)
Electrician wiring your house	1 Ω
Digital circuit designer	10 kΩ (10,000 ohms)
Linear circuit designer	500 kΩ
Special measuring equipment designers	10 MΩ (10,000,000 ohms)

Figure 4.5 What's considered a large value of resistance?

50 kΩ is considered large). Physicists and designers of special measuring equipment would consider 10 MΩ and, in special cases, even greater than 100 MΩ a large value. Resistors larger than 10 MΩ are not very common and resistors of 100 MΩ and larger are packaged in small glass tubes with the ends sealed. These large-valued resistors must be kept clean to prevent errors caused by current flow across the surface of the package.

The color code

Resistors have their values indicated by a color code. As shown in Figure 4.6, there are four color bands (sometimes only the first three are used) painted on the resistor body. By deciphering these bands (converting colors back into the numbers that the colors represent), the resistance value and the tolerance of the resistor can be determined. The bands are usually placed very close to one end of the resistor. The band closest to the end is the first number of the resistor value. The next band is the second number of the resistance value. The numerical value of the resistance is specified by these two numbers. The third band indicates the number of zeros, the scale factor. The fourth band, if used, indicates the tolerance.

An example of how a 100-Ω resistor is color coded is shown in Figure 4.7. The first band is brown (indicating 1) and the second band is black (indicating 0). The number of zeros needed for 100 is one (after the 10 from the first two bands), so the third band is also colored brown. Look at the resistors in the lab kit and make sure that the resistors have been mounted properly. Check to make sure that the resistor color

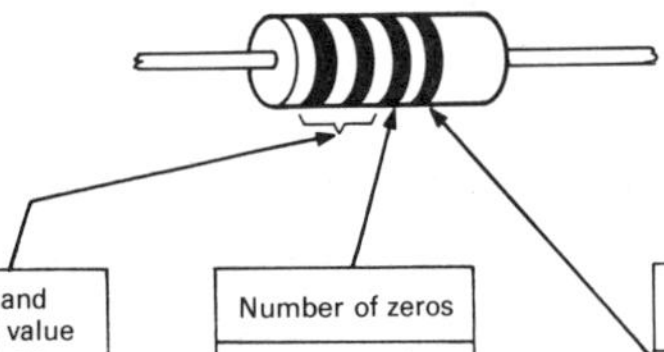

First and second value numbers	
Black	= 0
Brown	= 1
Red	= 2
Orange	= 3
Yellow	= 4
Green	= 5
Blue	= 6
Violet	= 7
Gray	= 8
White	= 9

Number of zeros	
Black	= 0
Brown	= 1
Red	= 2
Orange	= 3
Yellow	= 4
Green	= 5
Blue	= 6
Silver	= 0.01
Gold	= 0.1

Color of tolerance band	Resistor tolerance	Permitted range of a 100-Ω resistor
Silver	+ 10%	90-110
Gold	+ 5%	95-105
No marking	+ 20%	80-120

Examples:
Brown-black-brown = 100 Ω
Brown-black-red = 1000 Ω,1k
Brown-black-orange = 10,000 Ω,10k
Brown-black-yellow = 100,000 Ω,100k

(Note: This coloring is usually not used on wirewound resistors; the resistance value is printed on the body.)

Figure 4.6 Breaking the resistor color code.

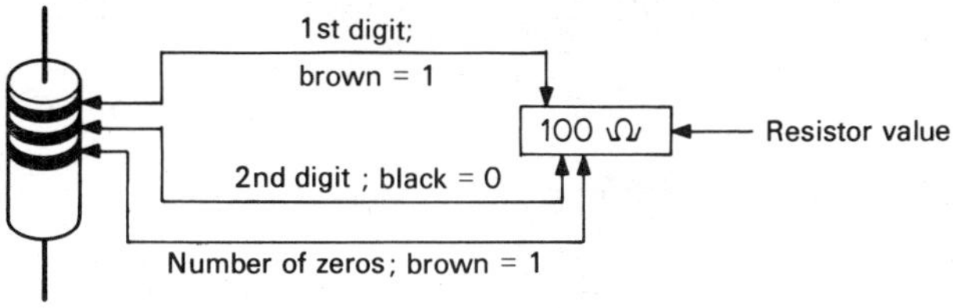

Check to make sure that all of the kit resistors indicate the same vulve as printed on the board under them.

Remember:	1	0	00	
1k =	Brown,	Black,	Red:	Red is the third stripe for resistors from 1k to 9.1k
	1	0	000	
10k =	Brown,	Black,	Orange:	Orange is the third stripe for resistors from 10 k to 91 k
	1	0	0000	
100k =	Brown,	Black,	Yellow:	Yellow is the third stripe for resistors from 100 k to 910 k

Figure 4.7 An example using the resistor color code.

code matches the printed value of resistance that appears next to each resistor on the component board of the lab kit. As an additional aid, remember that 1-kΩ resistors (or resistors in the range of 1 to 9.1 kΩ) have red as the third band. Resistors in the 10- to 91-k range have orange as the third band. Resistors in the 100- to 910-k range have yellow as the third band.

Measuring resistance

With this background on resistors, we will now set up the digital multimeter so that it can be used to measure the value of a resistor. This will convert the multimeter into an "ohmmeter." The black test lead stays in the second hole over from the left and the red lead is moved over to the left side, to the jack labeled "+ DC mA/kΩ," as shown in Figure 4.8. For ac or dc volts, the red lead is located to the right of the black lead; for ohmmeter functions or milliampmeter functions, the red lead is moved over to the left side. The meter jack on the left end is labeled "200 mA MAX." In the experiments that are shown in this book, values of current as large as 200 mA won't be encountered. The black lead will always stay in the same meter jack and the red lead will simply move back and forth to change from voltage readings to current or resistance readings. This meter will automatically provide the correct range (autorange) for any of these readings.

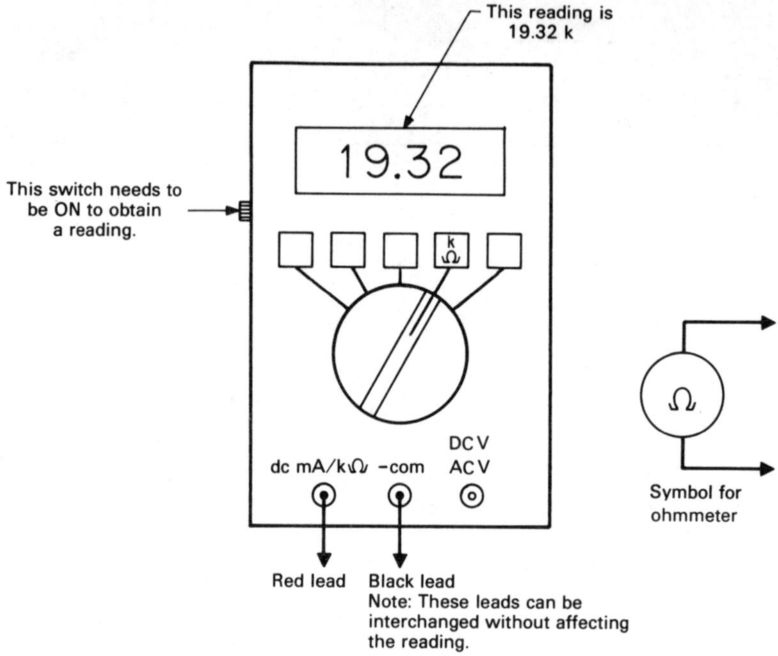

Figure 4.8 Setting up the digital multimeter to measure resistance.

There are five full-scale ranges for resistance measurements: 200 Ω and 2, 20, 200, and 2000 kΩ. If a resistor being measured is too large in value for the largest full-scale range, the leading 1 on the display will blink. This overrange indication also is displayed for an open circuit. A diode check feature is made available specifically to check semiconductor diodes to see if they are shorted, open, or okay, but it can't be used to check LEDs because the maximum voltage across the meter leads is maintained at 1.5 V and LEDs require larger voltages. This diode check function will not be used. Remember that the left-side power switch must be turned on to power the multimeter before making any of the readings.

When using the ohmmeter function, the red and black meter leads lose their significance. There is no concern with which end of a resistor goes to the red and which end goes to the black lead because there is no concern about polarity in the measurement of resistance.

The resistors in the lab kit are located in 4 columns of 5 rows each, for a total of 20 resistors, as shown in Figure 4.9. Each one of these resistors has the value printed next to it. It is a good exercise to read the color-coded value, then measure the actual value of that resistor, and write it down somewhere convenient. In some of the later experiments, in which the value of a particular resistor is involved in a cal-

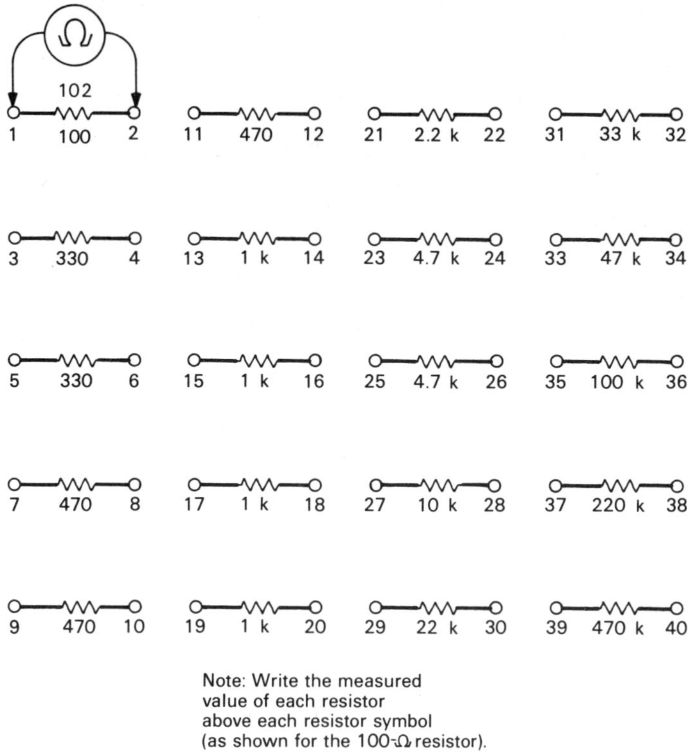

Figure 4.9 Measure and record the resistors in the lab kit.

culation, accuracy can be increased by using the actual measured resistance value rather than the marked value.

Read a few of these resistors to get the idea; read as many as needed in order to feel comfortable with the colors of a wide range of resistance values. It might be good to start with the 100-Ω resistor and then read a few kilohm resistors (like 1 k or 4.7 k), then 10 or 33 k, and finally 100 through 470 k. Read a few resistors in each of these relative sizes and use the meter to confirm the values (to also get an idea of how to use the multimeter as an ohmmeter).

Ohmmeters are often used as continuity indicators to determine whether a wire is still connected. This is useful in troubleshooting to test for failure of lamps, interconnecting wires, transformers, and coils. There are a lot of different transformers, coils, and other components in the lab kit, and Figure 4.10 shows the typical ohmmeter readings that you should obtain from the various terminal pairs that are associated with these components. If you attempt to measure the resistance of the earpiece, you will get an open circuit indication. The reason is that this earpiece is really a capacitor, a piezoelectric device. The elec-

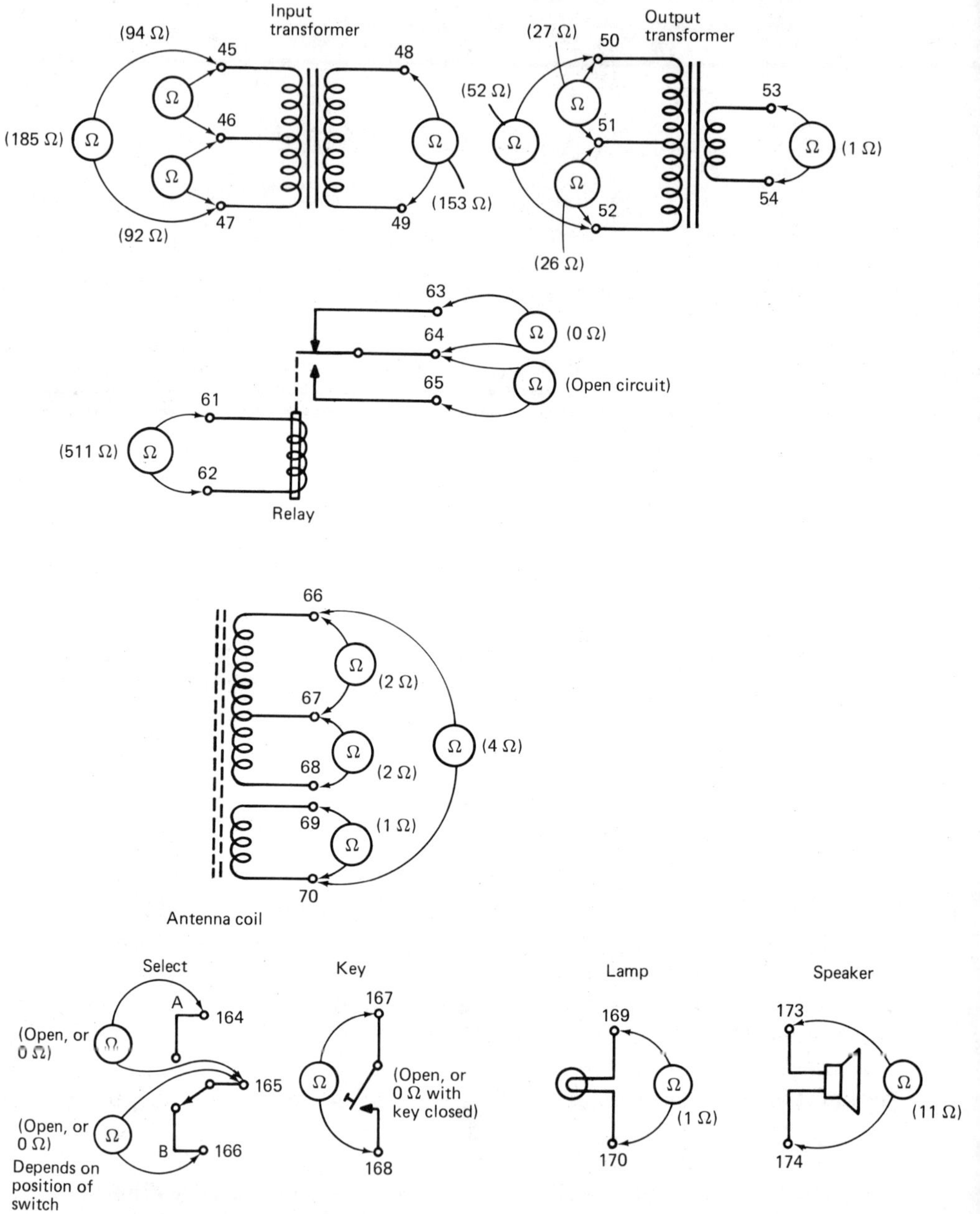

Figure 4.10 Measuring the resistance of other lab kit components.

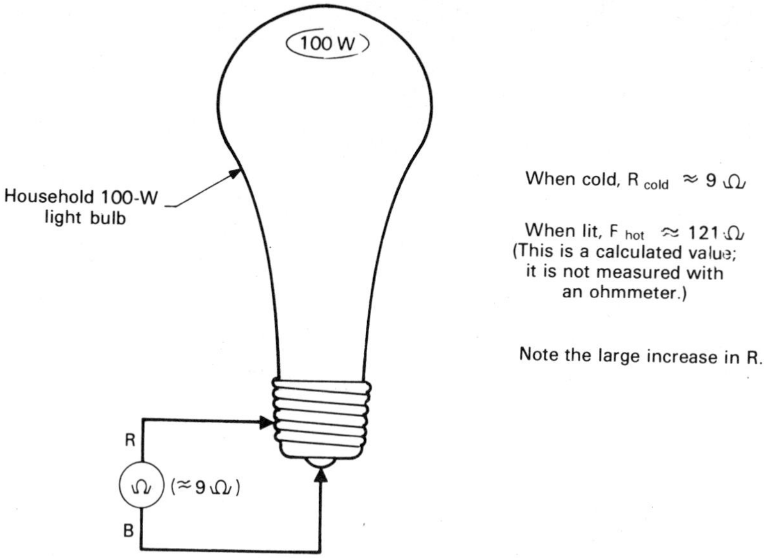

Figure 4.11 Measuring the cold resistance of a 100-W light bulb.

trical excitation actually mechanically strains (deforms) the crystal and creates sounds in this manner. (Another piezoelectric device, which we will describe later in the chapter, is the quartz crystal.)

When the measured resistance values become small (less than 5 Ω), subtract 0.1 Ω from the reading. If the test leads of the meter are shorted together, a reading of 0.1 Ω will be obtained. This resistance exists in the wires of (and the contacts to) the test leads. Therefore, to get higher accuracy for lower-valued resistors, subtract this internal 0.1 Ω from the value indicated by the meter.

The resistance of the filament of an incandescent lamp makes large changes from when it is cold to when it is turned on and becomes very hot. Measure the cold resistance of a 100-watt (W) light bulb by applying the terminals of the ohmmeter to the end metal contacts that exist on the light bulb, as shown in Figure 4.11. A typical reading would be about 9 Ω (an overrange indication would indicate a burned-out light bulb). When the lamp is hot, the resistance value increases to about 121 Ω (this value is calculated knowing the applied voltage and the wattage rating, as will be discussed later in this chapter). This is an increase of 13 times in resistance value. This low cold resistance of the filament allows a large in-rush of current when a light is first turned on (this is the reason that light bulbs usually burn out at this time).

On the front panel of the lab kit (just to the left of the loudspeaker, see Figure 4.12) is a cadmium-sulfide (Cd-S) photocell. This cell is

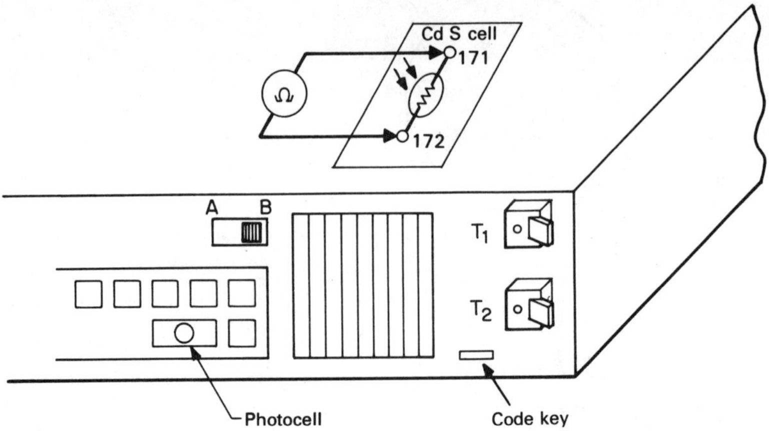

Cover up photocell so no light can fall on it and measure the "dark resistance."

Note the large change in resistance that is caused by light shinning on the photocell.

Figure 4.12 A photocell in which resistance depends on light.

exposed on the front panel (to the left of the loudspeaker) just above the large LED panel lamp and is available between terminals 171 and 172.

Cd-S cells are used in photographic light meters because this material drastically changes resistance value in the presence of light. Therefore, the Cd-S cell is a transducer that converts light intensity into a corresponding value of resistance. A circuit can be designed to respond to this variation in resistance and thereby respond to light intensity.

Resistors in series and parallel

Two 1-k resistors can be combined in either series or parallel as shown in Figure 4.13. With the two resistors in series (Figure 4.13*a*), the total resistance becomes the sum of the individual resistor values.

In Figure 4.13*b* the resistors are connected in parallel (also called a *shunt* connection). When the resistors are connected in parallel, the resulting total resistance becomes smaller than the value of either of the two resistors (as indicated by the equation shown in Figure 4.13*b*). This equation indicates that the equivalent resistance is the product of the two resistor values divided by the sum of the resistor values. Two 1-k resistors in parallel therefore have a total value of resistance of 500 Ω. If more than two resistors are connected in parallel, the resis-

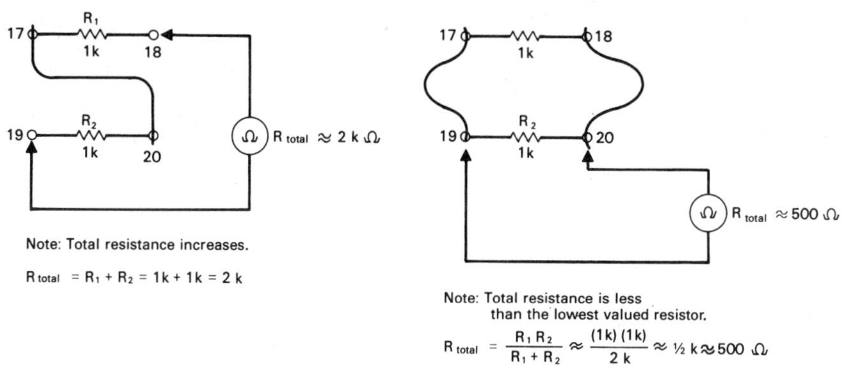

Figure 4.13 Combining resistors. (*a*) Resistors in series and (*b*) resistors in parallel.

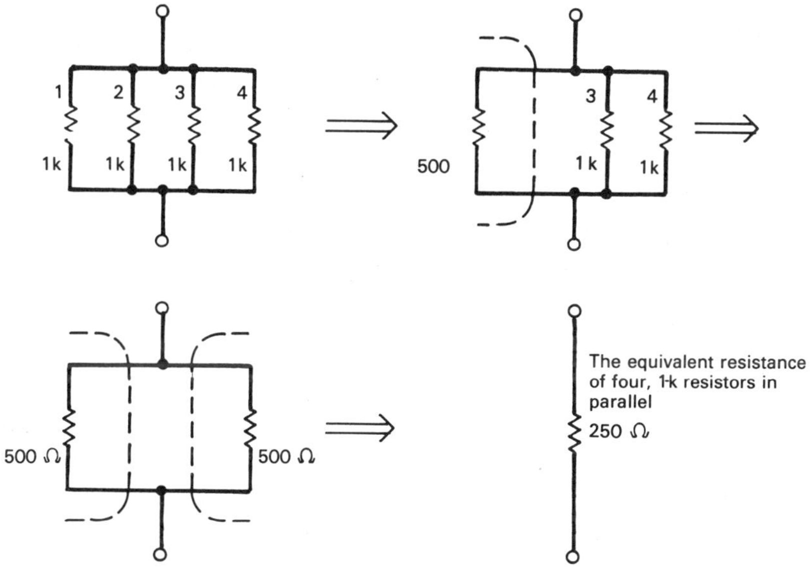

Figure 4.14 Finding the equivalent resistance of paralleled resistors.

tors can be combined (two at a time) into an "equivalent resistor," until the total equivalent resistance is found, as shown in Figure 4.14.

Some uses for resistors

There are many uses for resistors in electronic circuits. A very common application is to use two resistors in what is called a "resistive voltage divider." This is a way to reduce the magnitude of a voltage. For example, the 4.5-V tap from the battery section of the lab kit can

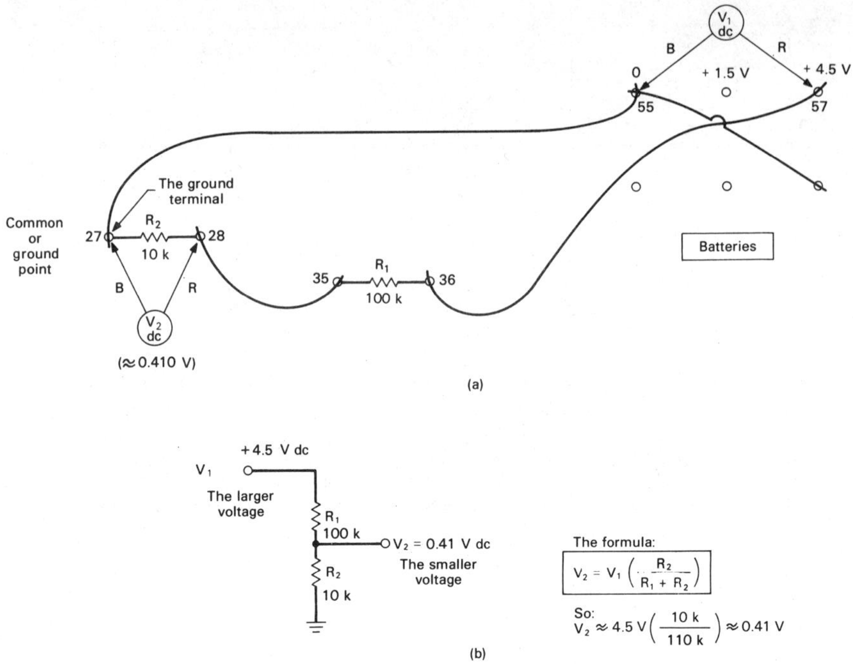

Figure 4.15 Resistors provide small voltages from large voltages. (*a*) A pictorial diagram and (*b*) a schematic diagram.

be reduced to 0.4 V by using two resistors with values of 100 kΩ and 10 kΩ, as shown in Figure 4.15

The equation shown in Figure 4.15*b* indicates how to calculate the voltage that results. Any two resistors that are available in the lab kit can be used and a calculation can be made to predict the resulting voltage. Also, any one of the voltages that are available from the battery pack can be applied to these resistors to gain some experience with this equation. Experiments like this will also provide a better feeling for resistive dividers (also called "resistive attenuators").

The potentiometer in the lab kit is labeled "Control" and has a front panel knob located to the left of center also labeled "Control." The total value of this resistance element can be measured between terminals 137 and 139, a value of about 50 kΩ, as shown in Figure 4.16*a*. To connect the potentiometer as a rheostat (Figure 4.16*b*), take a short wire and connect terminal 138 (the wiper) to terminal 139 (one end). The resistance (measured from terminals 137 to 139) will vary from approximately 0 Ω to 50 kΩ as the Control knob is rotated.

Figure 4.16*c* shows a voltage divider application of a potentiometer. In this case, notice that a 1.5-V cell is connected across the resistive

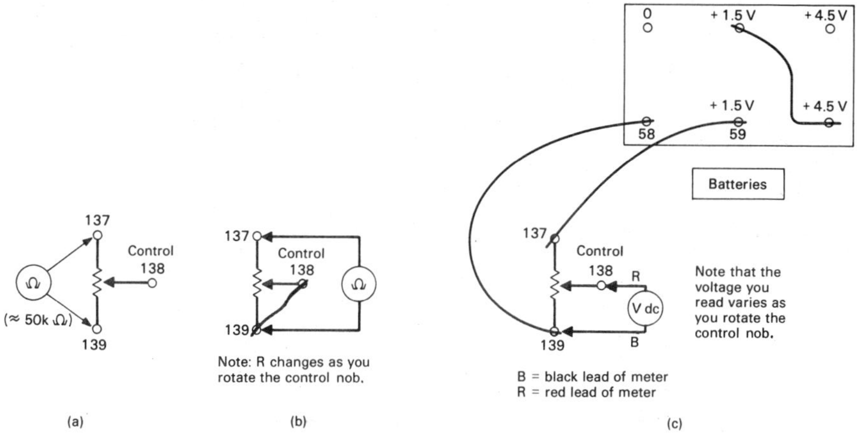

Figure 4.16 The rheostat and the potentiometer. (*a*) Measuring the total resistance, (*b*) rheostat connection, and (*c*) potentiometer connection.

element. Two wires (one going from battery terminal 58 to the potentiometer terminal 139 and a second wire going from battery terminal 59 to the potentiometer terminal 137) are used to connect 1.5 V across the potentiometer. If the multimeter is set up as a dc voltmeter (as previously used, where the red lead is moved to the right of the black lead to the jack in the lower right-hand corner, and the selector switch is moved to the red "DC V" position), the voltage values that the wiper picks up will be seen to vary from a maximum value of 1.5 V down to essentially 0 V as the Control knob is rotated. Any desired value of voltage can therefore be obtained. As the knob is turned clockwise, a larger voltage will be obtained. If the wires connecting to the resistive element of the potentiometer are reversed, the voltage readings in response to the direction of rotation of the knob will be reversed.

The potentiometer can be modified, as shown in Figure 4.17, by adding resistors at either or both ends. In this figure, a resistor is used at both ends so that the range of the voltage values that can be provided by the potentiometer can be controlled. Both a minimum and a maximum voltage value are established by these resistors. Any voltage that lies between these minimum and maximum values can be selected by the wiper.

A resistor and potentiometer made from common materials

To get a better intuitive feeling about resistors, a "carbon-film resistor" will be made. All it takes is a piece of paper and an ordinary lead pencil. In the corner of the piece of paper, draw a line about ¼ inch

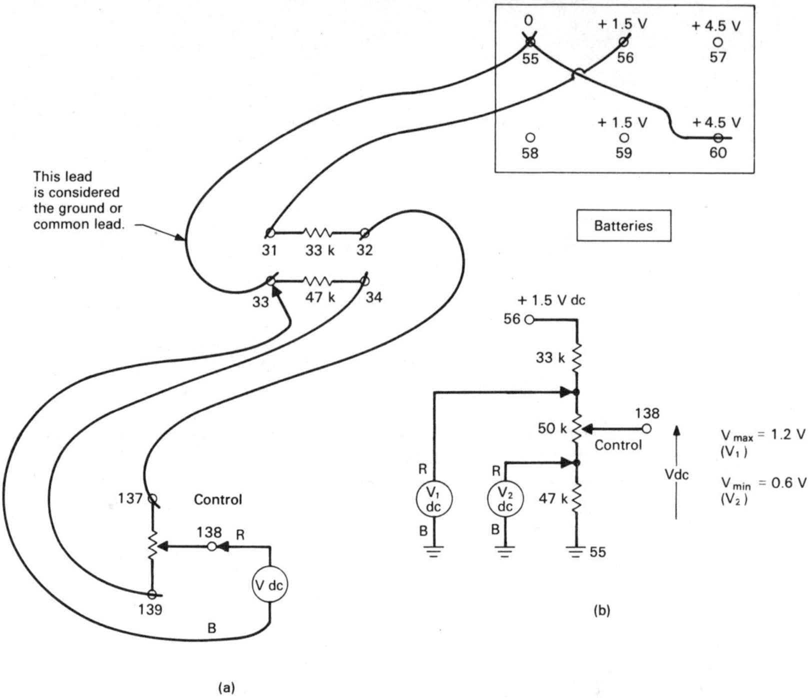

Figure 4.17 Using resistors to establish a limited voltage-adjustment range for a potentiometer. (*a*) Pictorial diagram of the circuit and (*b*) schematic diagram of the circuit.

(in) wide from one edge of the paper to the other, as shown in Figure 4.18. Fill this in by going back and forth with the pencil until the pencil line is a uniform dark color. On either end, where the line comes to the edges of the paper, enlarge the areas somewhat because that is where the wires will be connected to this homemade resistor. This pencil mark on the paper is a carbon-film resistor because pencil "lead" is made out of carbon (graphite).

As shown in Figure 4.19, paper clips are used to attach the ends of the longer wires that are available in the lab kit to the resistor element. These wires then go over to terminal posts T_1 and T_2 that are located on the front corner of the lab kit: the red and black terminal posts. This makes the resistor available via terminals 175 and 176. Set up the multimeter to read resistance and read the value of the resistance of this carbon-film resistor.

You will notice that it is a little bit of a problem to attach wires to the ends of this resistor. The attachment of wires is also one of the harder problems in the manufacture of resistors. Problems of these con-

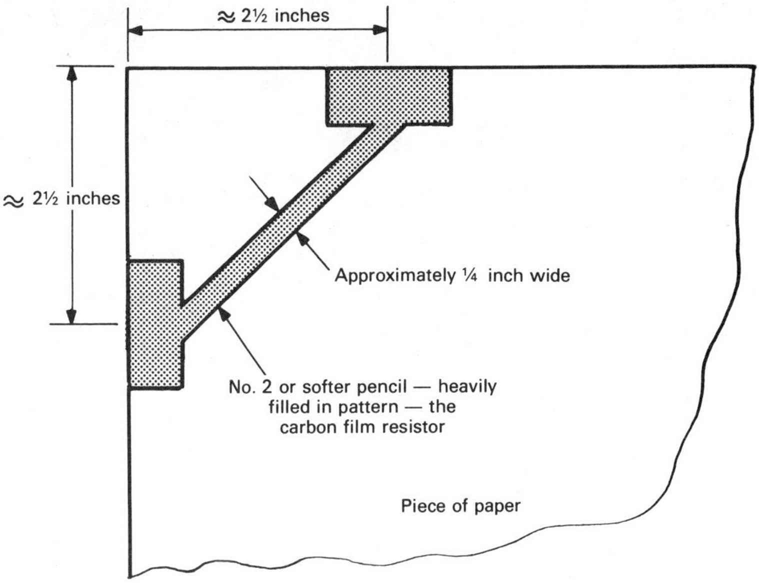

Figure 4.18 Making a carbon film resistor.

necting wires breaking loose or becoming intermittent is one of the largest reliability problems of resistors. You are therefore experiencing a lot of the problems that actually exist in the manufacture of resistors.

Take the resistor just made and apply a voltage of 1.5 V across the total resistance (using terminal posts 175 and 176). Now set up the multimeter as a dc voltmeter and run the red probe up and down the pencil trace, as shown in Figure 4.20. This is a potentiometer. The red probe is the wiper and the total resistive element is the carbon-film resistor. In this example, if the red probe is moved down to the lower-left end, the reading will be essentially 0 V. As the red probe is moved up, the voltage will increase smoothly until it reaches a maximum of 1.5 V at the upper end. Any intermediate voltage can be provided by properly locating the red probe.

Temperature effects on resistance

In linear, or analog, circuits, many times the precise value of a resistor is important for proper circuit operation. As a result, there is also concern with what happens to the value of a resistor if the temperature should change.

To obtain an idea of how temperature affects a carbon-film resistor, use a hair dryer and direct the heat onto the resistor, as shown in Figure 4.21. While the resistor element heats, watch the change in the

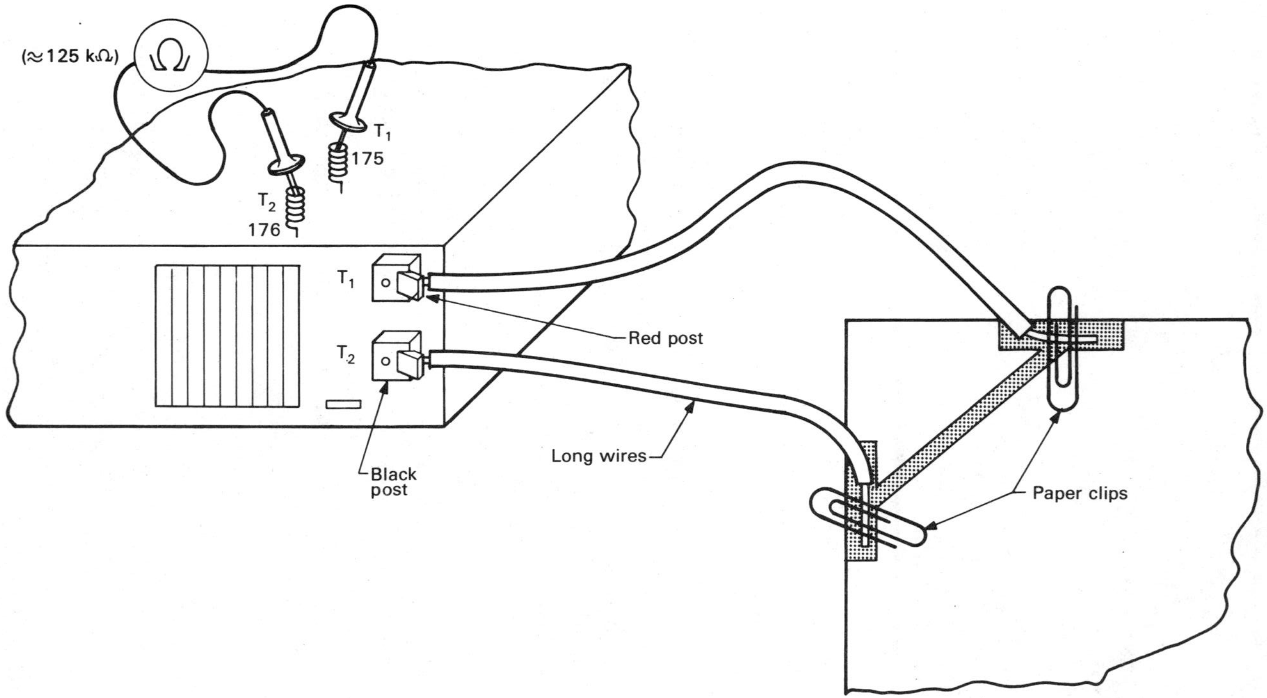

Figure 4.19 Attaching wires and measuring your carbon-film resistor.

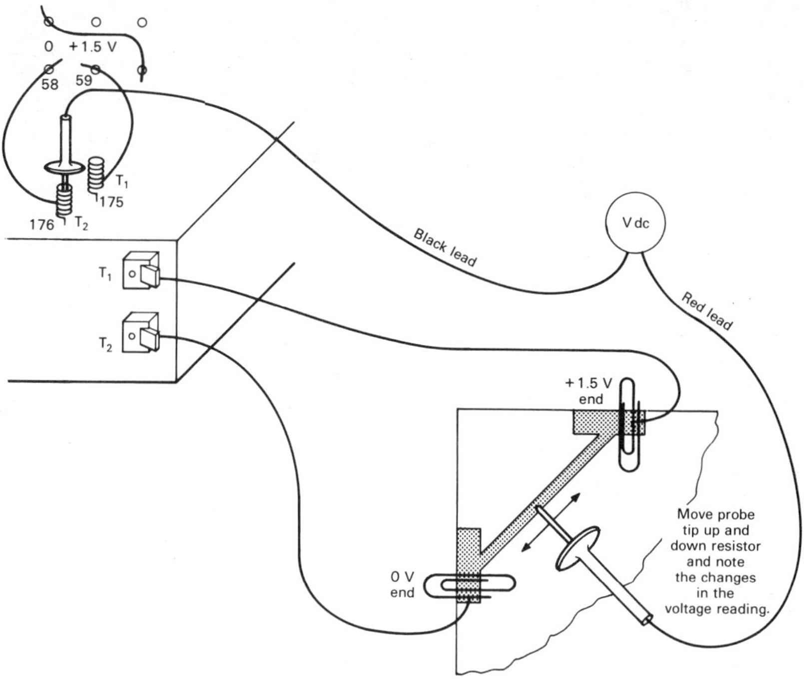

Figure 4.20 Making a potentiometer.

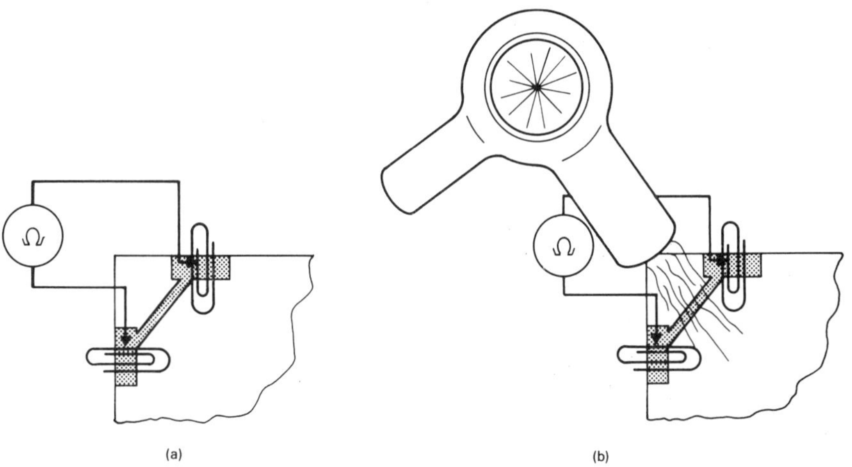

Figure 4.21 Determining the effects of temperature on your resistor. (*a*) Read and record resistance value at room temperature and (*b*) read and record resistance value at high temperature. Did R value increase or decrease?

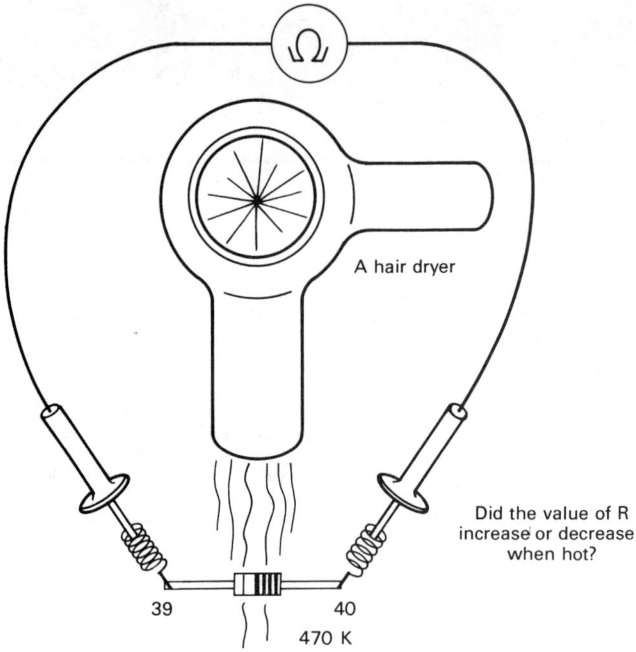

Figure 4.22 Determining the effects of temperature on a resistor from the lab kit.

resistance value that is indicated on the ohmmeter. The resistance value gets smaller when the resistor is heated. This is implying a negative change in the value of the resistor as the temperature is increased. This is called a "negative temperature coefficient." Carbon film is one of the few resistive materials that has a negative temperature coefficient. Most metals (and even most resistive materials) have a positive temperature coefficient.

The temperature effects on the resistors that are available in the lab kit can also be determined. Use the 470-kΩ resistor, as shown in Figure 4.22. Read the resistance value and then bring the hair dryer 2 to 3 in away from the resistor. Heat the resistor and watch the change in resistance that results. The resistance value will most likely decrease as temperature is increased because many of the resistors that are used in electronics are made out of carbon film.

Current: The Intensity of the Electron Flow

So far we have discussed voltage and resistance. The logical next subject to consider is current. The value of current or the amount of current flow is measured in units of amperes. One ampere is defined as 1 coulomb (C) of charge flow per second. This requires 6.24×10^{18} electrons (each electron has a negative charge of 1.602×10^{-19} C, the smallest possible amount of charge) to move by a fixed point every second to carry the charge flow equal to 1 A (ampere is usually abbreviated as A or amp). One picoamp requires a flow of only 6.24×10^{6} electrons per second.

The units for electrical and electronic parameters are usually named after famous scientists who, in the past, have made contributions to either the art of electricity or electronics. So resistance is measured in ohms in honor of the German scientist Georg Simon Ohm; voltages are measured in volts in honor of the Italian scientist Volta; current is measured in amperes in honor of the French physicist A. M. Ampere; and charge is measured in coulombs in honor of the French physicist Charles Augustin Coulomb.

What are large values of current? As shown in Figure 4.23, this again depends upon who is asked. An industrial electrician working around large factories would consider 500 A or larger high values of current. These currents would be going to the machinery that is used in a factory. Residential electricians, on the other hand, might think that currents larger than 150 A are large. An electronic service technician will think that 10 A is a large current. But, a digital integrated circuit designer would consider a large current to be something like 100 mA (one-tenth of an amp). Linear IC designers many times consider 10 mA a large current, especially if this is the current drain of a single IC circuit. In flashlights the typical current flow is 300–400 mA. In electronics we deal with a wide range of current flow, from perhaps tens of amperes clear on down to very small values (picoamperes).

Today, everyone tries to make electronic systems and electronic circuits that will operate with very small currents. This creates many advantages. If an electronic system is battery operated, long battery life results if only small current flow exists in the circuitry. Even if the electronic device plugs into a wall outlet, it is a benefit to only draw small currents. Then the electronic system doesn't get hot and therefore fans or other ways to cool the electronics are not needed.

Person	Currents larger than
Industrial electrician	500 a ac
Residential electrician	50 a ac
Electronic service technician	5 A dc
Digital IC designer	0.1 A (100 mA) dc
Linear IC designer	10 mA dc

Figure 4.23 What is considered a large value of current?

Measuring current flow

The multimeter can be set up to read dc current, as shown in Figure 4.24. The red lead stays in the same jack that was used for measuring resistance. The only change needed is to move the selector switch to point to the box labeled "DC mA." There is only one range for dc current, and this is 200 mA full scale.

The currents that flow in the seven-segment display (located in the front center of the lab kit) can now be measured. This display can be powered by connecting terminal 144 of the display to a 4.5-V tap in the battery pack, as shown in Figure 4.25. The black lead of the milliampmeter connects to terminal 55 of the battery pack, and the red lead can successively be contacted to terminals 145 through 151 to measure the values of dc current that light up each of the individual seven segments. If the red meter lead is simultaneously connected (using additional hook-up wires) to all of these terminals (145 through 151), all seven segments will be lit. This puts an 8 on the display. By simultaneously contacting to various groupings of these segments, the numbers 0 through 9 can be displayed.

This seven-segment display uses light-emitting diodes (LEDs). These displays were very popular in calculators and watches but have recently been replaced by LCDs (liquid crystal displays, like you see in the digital multimeter). The LCD requires a much smaller amount of current flow, which increases battery life in portable equipment.

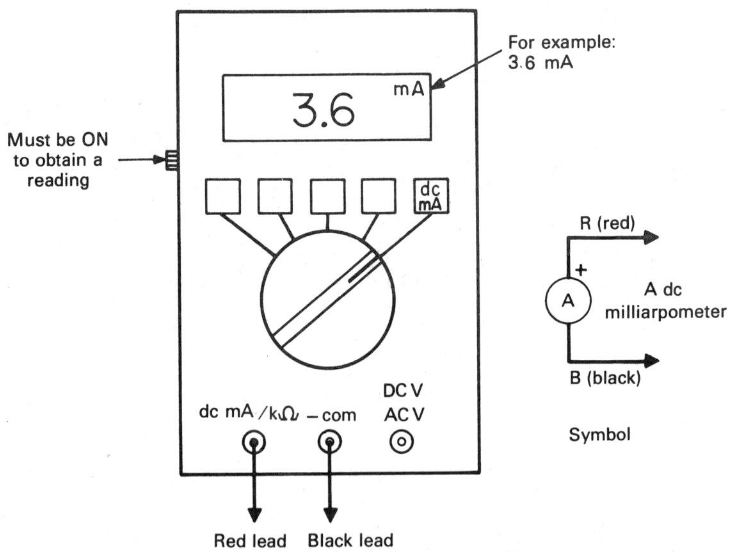

Figure 4.24 Setting up the digital multimeter to measure dc current.

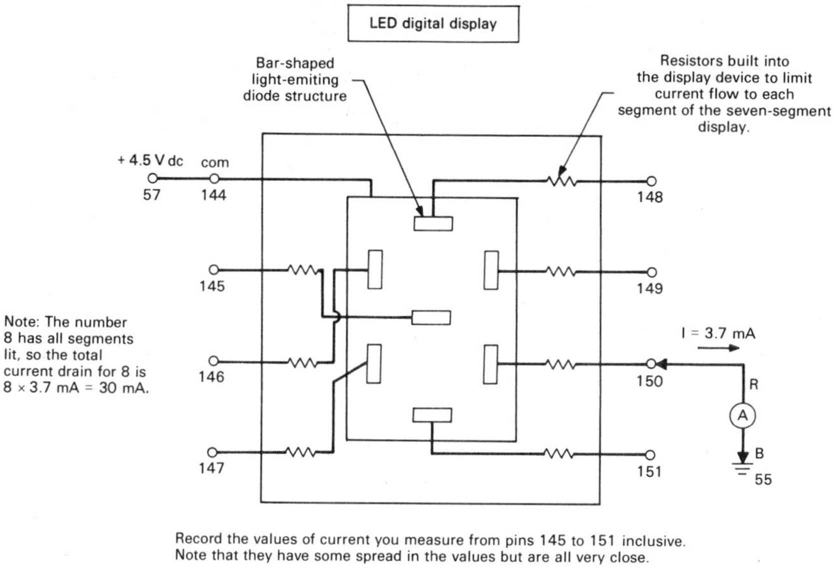

Figure 4.25 Measuring the currents used in a light-emitting diode display.

Six (LED) panel lamps exist above the seven-segment display on the front panel of the lab kit. These are simply panel-lamp indicators. They are providing an on/off indication, the same as an incandescent panel lamp. We will be experimenting with these shortly.

Ohm's law

There is a very important way to relate voltage, current, and resistance. This relationship is called "Ohm's law" in honor of Georg Simon Ohm. If a voltage (V) is developed across a resistance (R), this will specify the current (I) that is flowing ($V = IR$). Similarly, a current flow from a voltage source would specify the value of resistance that is connected to this voltage source ($R = V/I$). These examples of Ohm's law only apply for dc circuits (you will see that ac circuits are more difficult).

Relationships exist between the three parameters: voltage, current, and resistance. This gives rise to the three expressions of Ohm's law shown in Figure 4.26. The memory aid indicated here is to think of *voltage* as the *vulture* flying high in the sky, *resistance* as the *river*, and *current* (the symbol, *I*, "intensity," is used to denote current) as the *Indian*. The vulture flying in the sky looks down and sees the *Indian next to the river* (IR), $V = IR$. The Indian sees the *vulture over the river* (V/R), $I = V/R$. In a similar manner, the river "sees" the *vulture over*

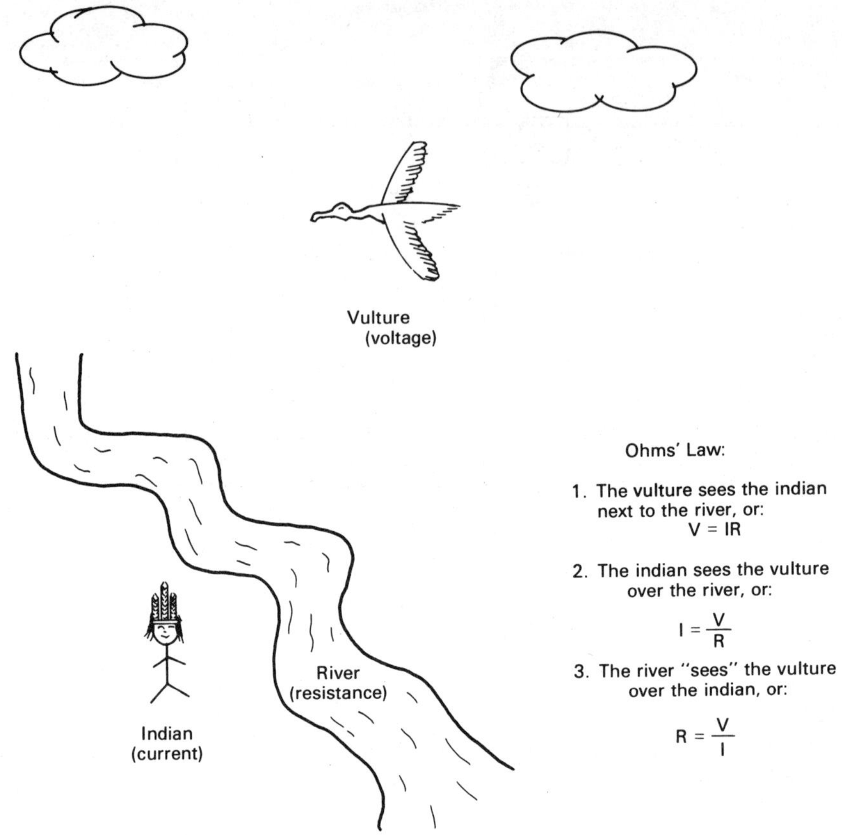

Figure 4.26 A memory aid for Ohm's law.

the Indian (*V*/*I*), $R = V/I$. By the use of the appropriate one of these expressions, any of these parameters can be calculated when the other two values are known.

Ohm's law can be used, as shown in Figure 4.27, to provide an indirect way to measure current flow. The multimeter has been previously hooked up as a milliampmeter and used to directly measure current, as shown in Figure 4.27*a*. Figure 4.27*b* shows an "indirect measuring technique." The multimeter is configured as a voltmeter, and the voltage that is developed across a current sampling resistor is measured. The current that must be flowing through this resistor can then be calculated using the relationship $I = V/R$.

In this example, the 1.5-V cell is used for the voltage source, so 1.5 V is across the 10-kΩ resistor. The calculation shown on this figure indicates that 1.5 mA of current will flow. This is called an indirect mea-

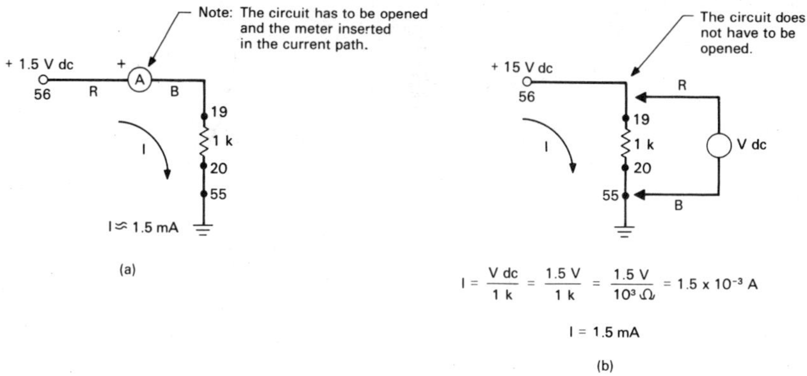

Figure 4.27 Two ways to measure current flow. (*a*) A direct measure and (*b*) an indirect measure.

sure because a voltage is actually measured, and then the current that flows is calculated. The advantage of this indirect technique is that voltages are easy to measure because the meter probe simply has to contact a circuit node. To directly measure current flow, the circuit has to be opened to insert the meter in series with the circuit. This is more difficult and takes more time. Therefore, this indirect way to measure current is often used.

Figure 4.28 shows that a 100-Ω resistor, when connected across a 1.5-V cell, will create 15 mA of current flow. This is a direct current measurement.

An interesting advantage of the indirect current-measurement technique can be demonstrated if the sampling resistor is changed from 100 Ω to 100 kΩ. The current can be calculated as 15 μA, as shown on the figure. The milliampmeter doesn't have enough sensitivity to directly read this small value of current. Smaller values of current can be measured with the indirect technique.

Resistors limit current flow

An example of the use of a resistor to limit current flow to an electronic component is shown in Figure 4.29. Here the 10-kΩ resistor is limiting the current that is flowing through one of the light-emitting diodes (LEDs) that is in the lab kit. The front panel LED labeled No. 1 (available on terminals 153 and 152) is used. The current can be determined by using an indirect current measurement. The value of voltage (V_2) that exists across the 10-k resistor is measured, and then the current is calculated (as shown in the figure) to be 318 μA. The light output from the LED can still be seen even though this small current is flowing (LEDs are usually operated at 10 mA). With optimum view-

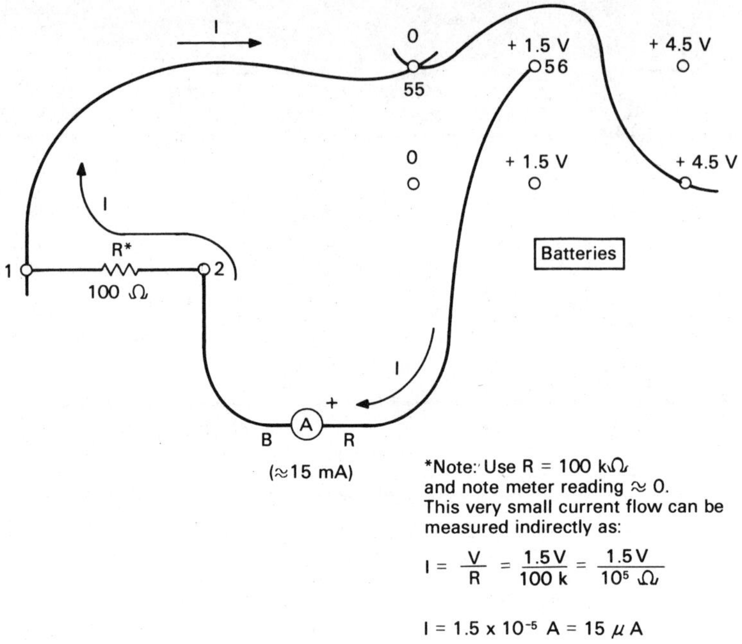

Figure 4.28 Resistors resist current flow.

ing conditions, the light output from an LED can be seen at much smaller values of current flow. Increase the resistor value to 100 k and see if light can be detected from the LED. Also, change the resistor to 330 Ω to see a brighter light output.

Voltage sources and current flow

In the previous chapter, "ideal voltage sources," as shown in Figure 4.30*a,* were assumed. An ideal voltage source will provide any amount of current that is demanded by the circuitry that is connected to it. If a small value of load resistance is connected across an ideal voltage source, a large current will flow, *but the output voltage will remain unchanged.*

Any real voltage source has a "source resistance," a small amount of internal resistance. Therefore the output voltage that is available to a load will fall somewhat because of the voltage drop that results as the load current flows through the source resistance. Larger values of load current create a larger drop in the output voltage.

An example of a voltage source that is *almost* ideal is the 110-V wall outlet in homes. A lot of different electrical appliances can be powered from a wall outlet and the voltage will still remain at 110 V. However,

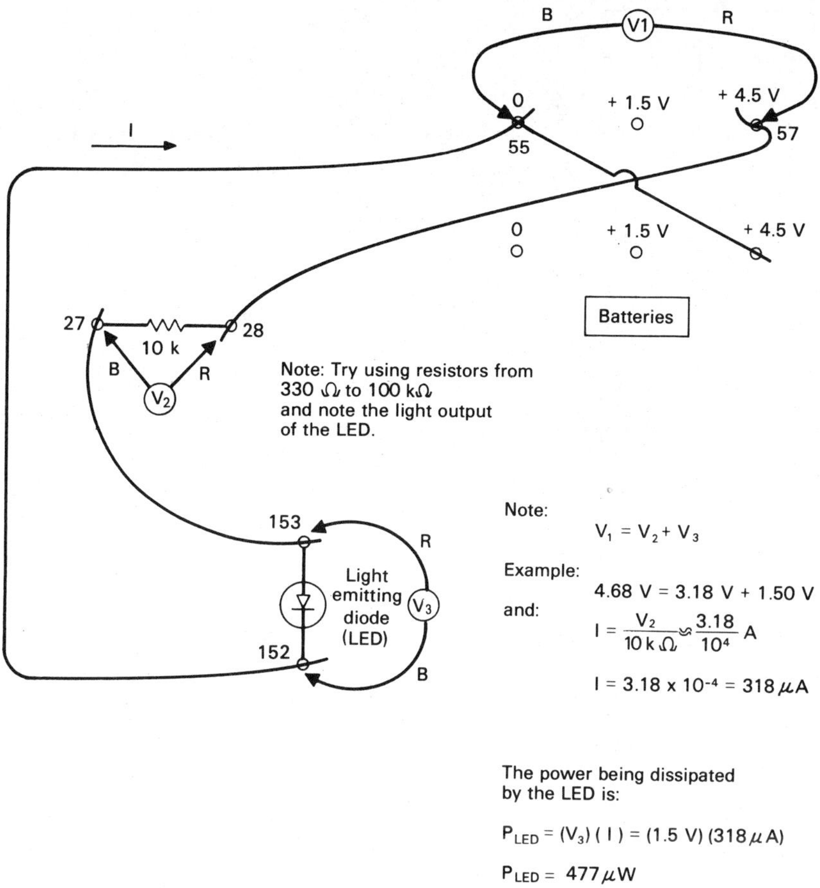

Figure 4.29 Resistors limit current flow to other components.

a radial-arm saw (or any other appliance with a large initial current demand) will cause lights to dim until the saw comes up to speed. Therefore the ac wall outlet is not an ideal voltage source because the source resistance, although small, does cause the voltage to drop when large load currents are demanded.

A problem can result from this source resistance if a motor is operated at the end of a long extension cord. Extension cords raise the source resistance of the ac power line, and if a motor can't get sufficient current from the extension cord, it will not start and can burn out. Lights, on the other hand, usually require less current and will simply reduce their brilliance if the source resistance of the power line is too large.

An example of a more ideal voltage source is a car battery. In fact, a simple way to test a car battery is to turn on the headlights before the

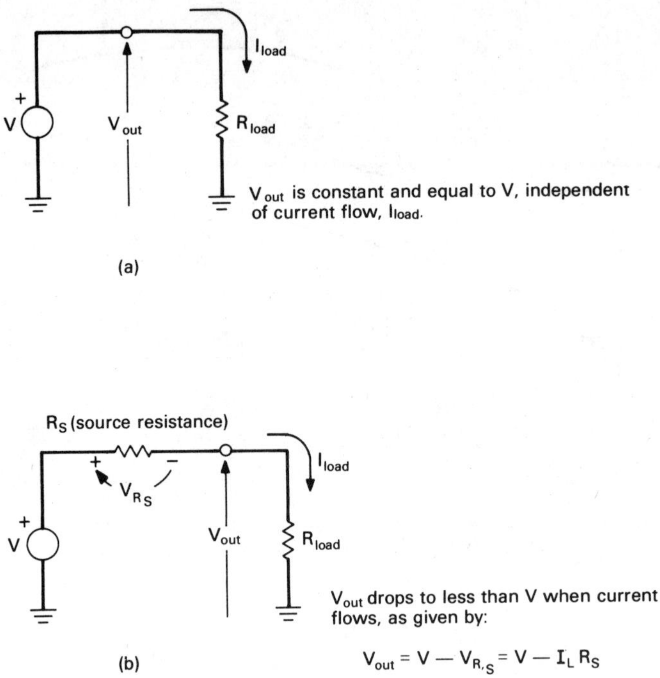

Figure 4.30 Real voltage sources have a source resistance. (*a*) An ideal voltage source and (*b*) a real voltage source.

car is started. The starter on an automobile draws a large amount of current from the battery. By observing the brilliance of the headlights while the starter motor is engaged, you can determine the condition of the battery. A good battery will provide essentially a constant headlight illumination even though the starter motor is engaged. A bad battery, on the other hand, causes the headlight intensity to decrease (or even change to a yellow shade) while the starter is engaged. Also, the starter motor will not turn the engine over very rapidly. The starter won't crank the engine rapidly because the terminal voltage of the battery falls when high currents are supplied from a bad battery. The reason for this is shown in Figure 4.30*b*.

Power: Whenever a Current Flows

When a voltage source causes a current flow, power is being delivered to a circuit, as shown in Figure 4.31. In Figure 4.31*b*, when the switch is closed and current flows, electrical power is being provided to the lamp. This power, in this dc circuit example, is the product of the voltage times the current. The units for electrical power are watts (W),

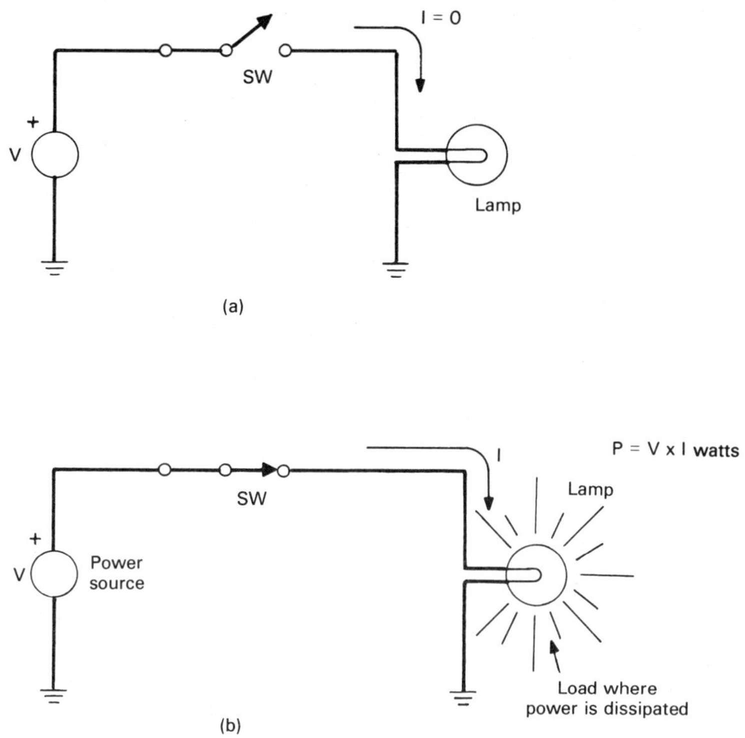

Figure 4.31 Power requires a current flow. (*a*) No power with $I = O$ and (*b*) if I, then P.

named in honor of the American inventor James Watt. To allow comparisons with other ways of obtaining power, 1 horsepower (hp) is equivalent to 746 W.

All resistors dissipate power when there is a voltage across them, and, because of this, they get hot (although with small power dissipation there may be only a very slight temperature rise). Therefore resistors are provided in many different packages in order to increase the power that can be dissipated. Figure 4.32 shows the smallest-size resistor that is commonly available, and this has a power dissipation rating of one-eighth of a watt. The power dissipation rating for a resistor is an important specification and ranges up to 2 W for the largest-size resistor shown in Figure 4.32 (although resistors with much higher power ratings are also available).

The dc power dissipated in a resistor is the product of the voltage across the resistor times the current flow through the resistor. This power can also be calculated as the square of the voltage across the resistor, divided by the resistance value, assuming the voltage and the

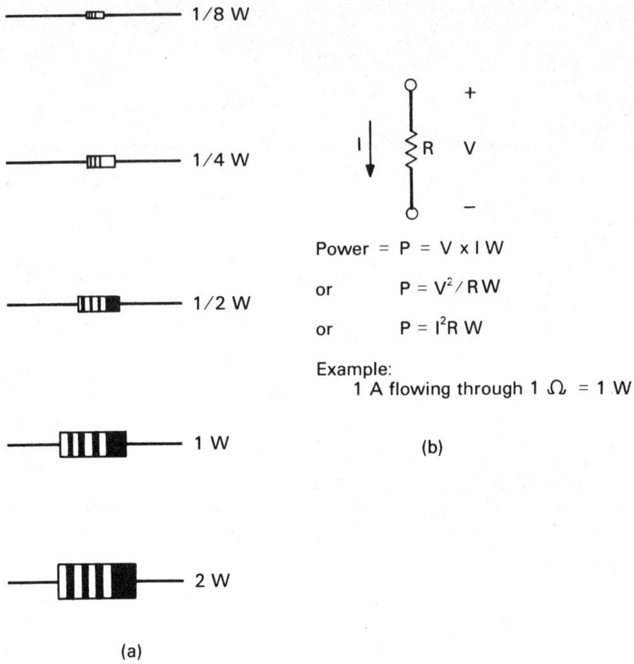

Figure 4.32 Resistors dissipate power: They get hot. (*a*) The larger the physical size, the larger the power that can be dissipated and (*b*) power is dissipated when current flows through a resistor.

resistance were known, not by the current flow. If we knew the current flow and the resistance, power would then be calculated as the square of the current times the resistance. This means, for example, that 1 A of current flowing through 1 Ω of resistance generates 1 W of power dissipation. Power is only dissipated when current flows.

Electrical energy and your electric bill

When you pay for the electrical energy that you use, the energy unit is the "kilowatthour." Electrical energy is the power that is consumed multiplied by the time during which this power is used. One kilowatt of power consumed for 1 hour (h) is 1 kilowatthour (kWh) of electrical energy. Also, 10 kW of power consumed for one-tenth of 1 h is 1 kWh.

In science, the unit for energy is the joule (J), and 1 J is 1 W\s. Therefore, 1 W being consumed for 1 s represents 1 J of electrical energy. There are 3.6 million J in 1 kWh.

The cost of electrical energy ranges from 5 to 10 cents per kilowatt hour (depending on locality), and the cost also increases as more power is consumed. An idea of what it costs to operate home appliances can

Assume the cost of electric energy = 10¢ / kW•h.

Example: What is the energy cost for a small 1200-W bathroom heater if operated for 25 min?

Answer:

$$\text{Cost} = 1200\ \text{W} \times 25\ \text{min} \times \frac{\$0.10}{\text{kW}\cdot\text{h}}$$

$$= \left(1200\ \text{W} \times \frac{1\ \text{kW}}{1000\ \text{W}}\right)\left(25\ \text{min} \times \frac{1\ \text{h}}{60\ \text{min}}\right)\left(\frac{\$0.01}{\text{kW}\cdot\text{h}}\right) = \$0.005 = ½¢$$

Figure 4.33 Determining the cost of operating an electrical appliance.

be determined. For example, the calculation shown in Figure 4.33 illustrates that it costs ½ cent to operate a small 1200-W bathroom heater for 25 minutes (min) if the rate is 10 cents per kilowatthour.

The electrical power drain of home appliances

It is interesting to consider the relative costs for operating electrical appliances. Pay attention to the high-power-drain appliances. Some of the smaller appliances don't really matter much one way or the other. Radios, for example, typically draw very small amounts of power; therefore they won't affect your electrical bill in a very dramatic fashion.

Figure 4.34 shows the range of the power drain, in watts or kilowatts (1000 W), of some of the common home appliances. There is a wide range in the power required. Neon night-lights (that glow with a yellow light) will consume only one-quarter of a watt, whereas incandescent night-lights consume 4 or 7½ W. A total of 400 neon night-lights that are on simultaneously will draw the same power as one 100-

Appliance	Energy use (kwh/month unless otherwise stated)
Electric heating	
Small home	350-700
Large home	700-2000
Hot water heater	480
Water bed heater	80-120
Central air cond. (3 T)	570-1705
Refrigerator (19 cu ft)	108
Household lighting	100
Radio-phonograph	0.03 kwh/h
Color tv	0.23 kwh/h
Washer, hot wash, warm rinse	0.25 kwh/load
Clock	2
Electric blanket (king)	1 kwh/night
Night light (4 W)	3
Vacuum cleaner	0.75 kwh/h

Figure 4.34 Electrical power requirements of some home appliances.

W incandescent light bulb. Some appliances consume kilowatts of power. Operating these appliances for extended time intervals rapidly raises the electric bill.

One of the largest electrical energy drains is the central home refrigeration units that are used to cool a house in the summer months. People who live in the desert regions and operate the central refrigeration unit night and day can run up electrical energy bills of several hundreds of dollars during the summer months. In this case, it becomes very important to keep the outside doors closed as much as possible and to consider other steps, such as increased insulation, to reduce the operating time and therefore the monthly costs of the central refrigeration unit.

By way of contrast with the large power-consuming appliances in homes, the digital multimeter used in this book consumes 5 milliwatts (mW). This means that only 1.7 mA is taken from the internal 3-V battery while the meter is on.

Capacitors: The Capacity to Store Electrons

The symbol for a fixed capacitor is two parallel lines separated by an air space, as shown in Figure 4.35. This is basically a diagram of how a capacitor is made; two plates of metal are separated by some kind of a nonconductor that is called a "dielectric" material. A "variable capacitor," shown in Figure 4.35*b*, uses essentially the same symbol as a fixed capacitor but adds an arrow going through the capacitor symbol at an angle. The curved line of this capacitor symbol is used to indicate the movable plates. Variable capacitors provide the station tuning control for AM and FM radios. When the knob is turned to tune to a different station, the value of a variable capacitor is typically changed. The knob marked "Tuning," located on the extreme left edge of the front panel, is the control for a variable capacitor (terminals 135 and

The unit of capacitance is the farad (F, a very large amount of capacitance).

More common units are: 1 microfarad = 0.000, 001 F = 1 μF = 10^{-6} F
1 nanofarad = 0.000,000,001 F = 1 nF = 10^{-9} F
1 picofarad = 0.000,000,000,001 F = 1 pF = 10^{-12} F

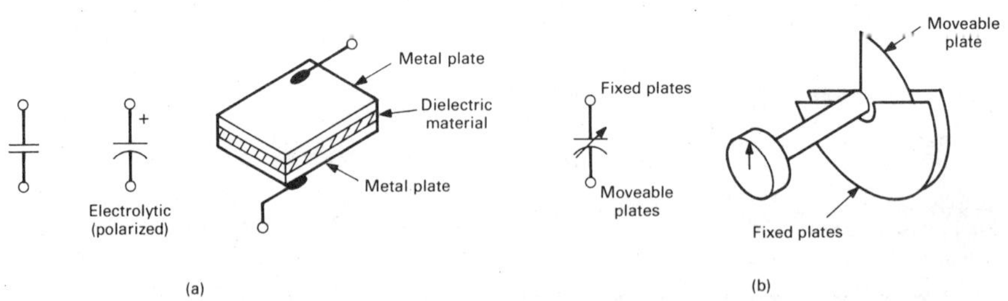

Figure 4.35 Capacitor symbols. (*a*) Fixed capacitors and (*b*) a variable capacitor.

136) that has a maximum value of about 265 pF. We will be using this variable capacitor to make an AM radio transmitter in Chapter 6. By simply adjusting this Tuning control, you will be able to change the transmitted frequency so that your transmitter will come in on your AM radio on any particular frequency that you would like.

A wide range of capacitors is shown in Figure 4.36, where it can be seen that many different physical types of capacitors are available.

Capacitors have the basic circuit property that they can store charge and, as a result of this, they tend to maintain a constant voltage across their terminals. A capacitor can accept current from a circuit and relatively slowly "charge up" to a new voltage value. A capacitor can also supply current to a circuit. This "discharging current" also tends to hold the voltage across the capacitor relatively constant. One of the circuit uses for a capacitor, therefore, is to make use of this tendency to resist voltage changes. Large-valued capacitors can act similarly to a battery. An example of this is demonstrated when the plug that was powering a radio is pulled out of the ac wall outlet: the radio will con-

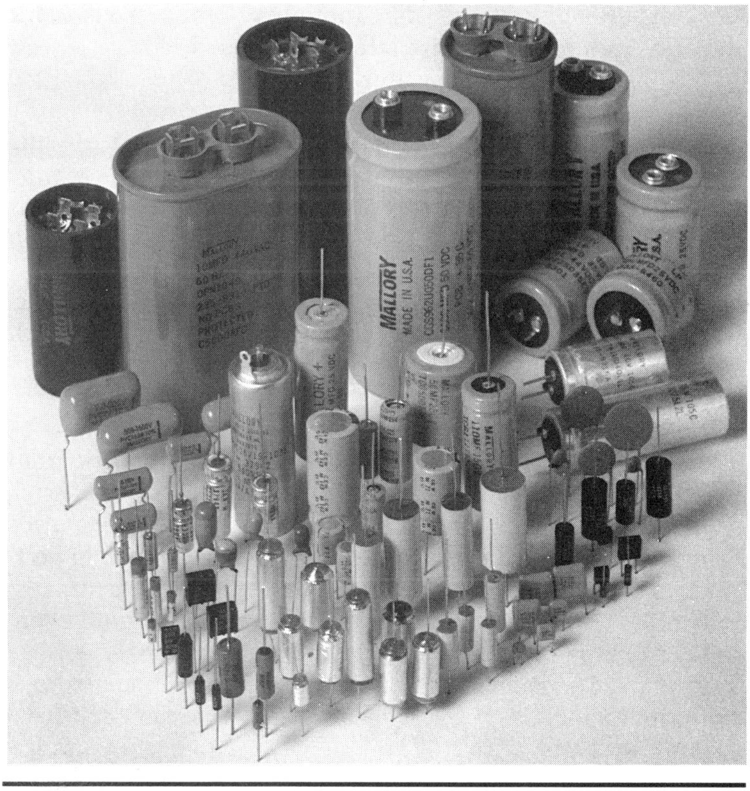

Figure 4.36 Capacitors look like this. *(Courtesy of Mallory Capacitor Company.)*

tinue to play for a short time. The radio is being powered by internal capacitors that slowly lose their charge. Also, unavoidable "parasitic circuit capacitance" undesirably limits the processing speed of the circuits within a computer because these "stray" capacitors tend to slow down the voltage changes that occur within the circuits.

Discrete component capacitors are supplied in a wide range of values. The basic unit of capacitance is the farad, named in honor of Michael Faraday. One farad (F) is an extremely large value of capacitance. Metric prefixes are used for the smaller capacitor values that are used in electronic circuits. More typical are capacitor values that make use of the following metric prefixes: microfarad (μF), one-millionth part of a farad (10^{-6} F); nanofarad (nF, 10^{-9} F); and picofarad (pF, 10^{-12} F). The abbreviation for picofarad, pF, is often pronounced "puff," as a "10-puff capacitor."

It takes a relatively large chip area to provide what is generally considered a small-valued capacitor within an integrated circuit. A large capacitance value is something like 50 to 100 pF, but in the real world of electronics, 100 pF is considered to be a very small value of capacitance. In contrast to this, resistors can more easily be supplied within integrated circuits. The move to integrated circuits has therefore greatly reduced the number of discrete resistors that are required in electronic systems, but external capacitors are often still needed.

In electronics, small-valued capacitors have values less than 1000 pF (also sometimes designated as 0.001 μF or more properly 1 nF). Medium-valued capacitors have values of capacitance up to 1 μF, and large-valued capacitors have capacitance values in excess of 1 μF.

Relatively large-valued "electrolytic capacitors" are available. These capacitors make use of internal electrolytic action to form a very thin "dielectric film" (an insulating oxide layer) on a metal surface. This thin film of oxide on one of the metal plates of the capacitor allows a very large magnitude of capacitance to fit into a small package. The other plate of an electrolytic capacitor is an electrolyte that is either wet or at least damp to provide an electrical conductor that is next to the thin oxide layer. DC voltage must be applied to an electrolytic capacitor to match the polarity marks that are indicated on the capacitor. A plus (+) sign is associated with one of the terminals of an electrolytic capacitor. The voltage applied to this terminal has to be always kept more positive than the voltage at the other terminal; otherwise the oxide film will be destroyed, and the capacitor will be ruined. It will lose capacitance and eventually conduct dc current.

There are ten fixed capacitors as shown in Figure 4.37, and one variable capacitor in the lab kit. These capacitors are located near the center of the kit. To the right, within the group of capacitors, are four electrolytic capacitors. Notice that these use a polarity indication. A plus

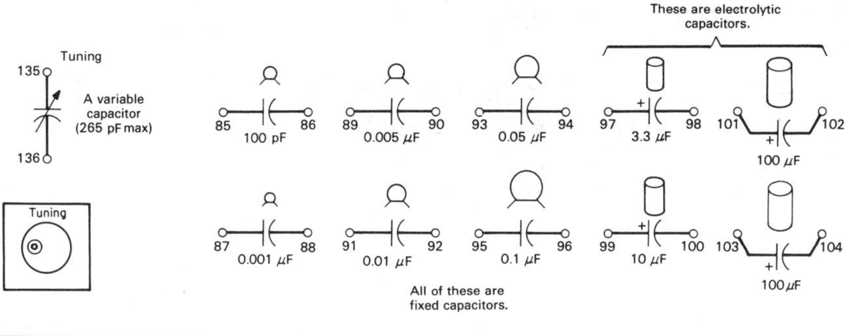

Figure 4.37 The capacitors in the lab kit.

sign appears next to the appropriate terminal for all four of these electrolytic capacitors. The six smaller-valued capacitors are all called "disc ceramics." The ceramic dielectric is a special material that provides large values of capacitance in a relatively small size.

Capacitors in series and parallel

Capacitors can be connected together. Figure 4.38 shows two capacitors connected in series and also in parallel. Capacitors that are connected in series provide a smaller total capacitance (the same as resistors connected in parallel). The total capacitance (C_T) of a series connection of two capacitors is the product of the capacitance values over the sum of these values. In contrast to this, if the two capacitors are connected in parallel, the total value of the resulting capacitance is the sum of the capacitor values (like resistors connected in series).

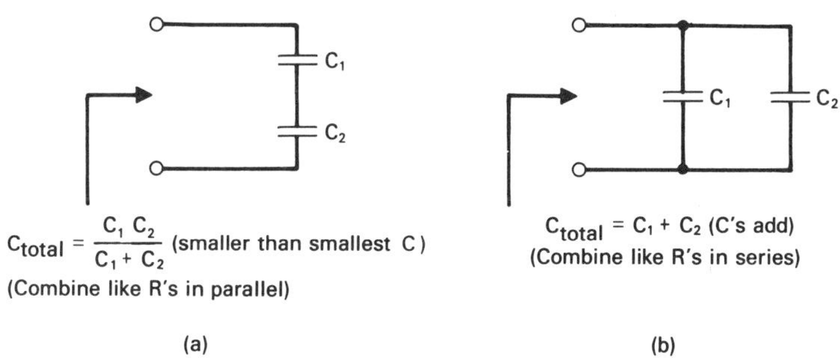

Figure 4.38 Capacitors in series and parallel. (*a*) Series connection and (*b*) parallel (shunt) connection.

Comparing an ideal capacitor with a real capacitor

An ideal capacitor does not dissipate any power, as indicated in Figure 4.39. Capacitors will accept a charge and can then deliver this charge back to the circuit at a later time. Real capacitors have some small losses associated with the dielectric material (or with the resistance of the wire connecting leads), so they will dissipate very slight amounts of power. Capacitors can heat up as the frequency of the applied voltage is increased because the dielectric losses of a capacitor often increase with frequency. A power dissipation rating is not used for a capacitor. The two critical specifications of a capacitor are the breakdown voltage and the capacitance value.

There are many additional specifications for capacitors, such as the leakage current (dc current that flows through the dielectric material). An ideal dielectric will not pass dc current. Real capacitor dielectrics sometimes do allow small amounts of dc current flow. If the capacitor is placed across the voltage source that is used to power a circuit, this small dc leakage current flow is not a problem. In some circuit applications, the dc leakage current of a capacitor is a major problem. Additional specifications indicate the time and temperature stability of the capacitance value.

Dielectric constant

In its simplest form, a capacitor has two metal plates that are separated by a dielectric. A dielectric is any material that does not conduct current, an insulating material such as air, mica, glass, plastic film, and special ceramic compounds such as barium titanate. The particular dielectric material that is used affects the value of the capacitance that can be obtained in a limited-sized package, and this dielectric also determines most of the other characteristics of the capacitor.

The value of a parallel plate capacitor with an air dielectric, Figure 4.40*a*, can be increased if a solid dielectric material is used, Figure 4.40*b*. This increase is caused by a characteristic of the dielectric mate-

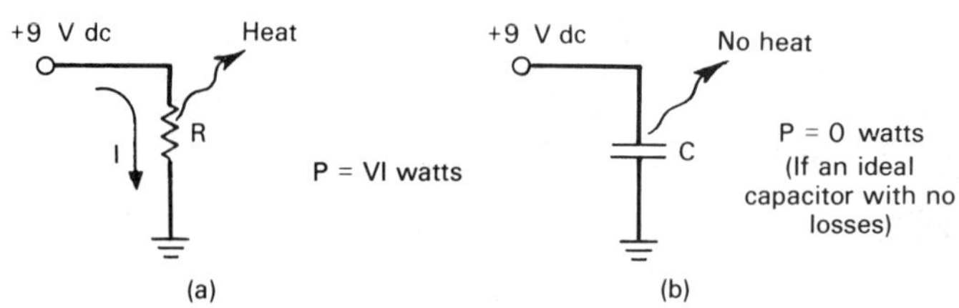

Figure 4.39 Capacitors are energy efficient. (*a*) Resistors get hot and (*b*) capacitors stay cool.

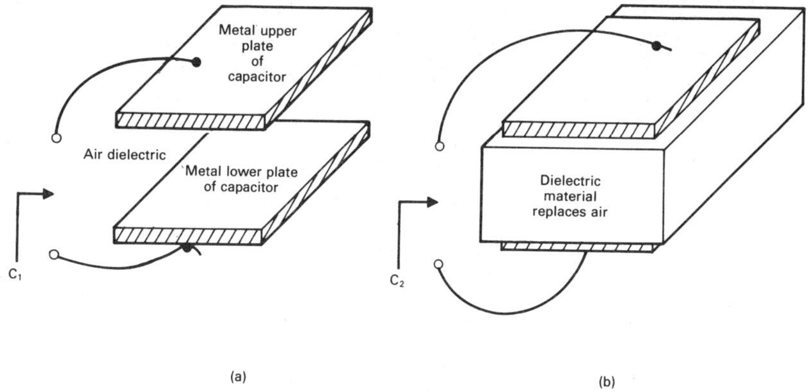

Figure 4.40 The meaning of dielectric constant. (*a*) Capacitance value with air as the dielectric and (*b*) capacitance value with dielectric material replacing the air.

rial called the "dielectric constant." For example, if the value of capacitance doubles, the dielectric constant of this material is 2. Paper has a dielectric constant of 2 to 3. A lot of plastics have dielectric constants in the 3 to 4 range. Some of the special ceramic dielectrics that are used in disc ceramic capacitors have values of dielectric constant of 1000 or even 10,000! These are known as "super K" (super dielectric constant) materials. A high dielectric constant material is desirable because a large value of capacitance can be obtained in a small-sized package. Unfortunately, other characteristics of these capacitors suffer, which limits their range of applications.

The voltage rating of a capacitor

The voltage rating of a capacitor is the most important specification. The next consideration is the value of capacitance. A circuit designer must ensure that the voltages that are applied across a capacitor are always less than the "breakdown voltage rating" of the capacitor.

Electrical breakdown results from the very intense electrical force that tries to reunite positive charges with negative charges. An "electric field" is said to exist within the region between separated groupings of charge. An electric field therefore results between the plates of a capacitor when a voltage exists between these plates, as shown in Figure 4.41. The magnitude of this electric field is given by the magnitude of the voltage divided by the distance that separates the plates. The units for electric field strength are volts per meter. If the electric field gets too high, the dielectric material will undergo an electrical breakdown. Many times this breakdown creates a spark and also a cracking noise.

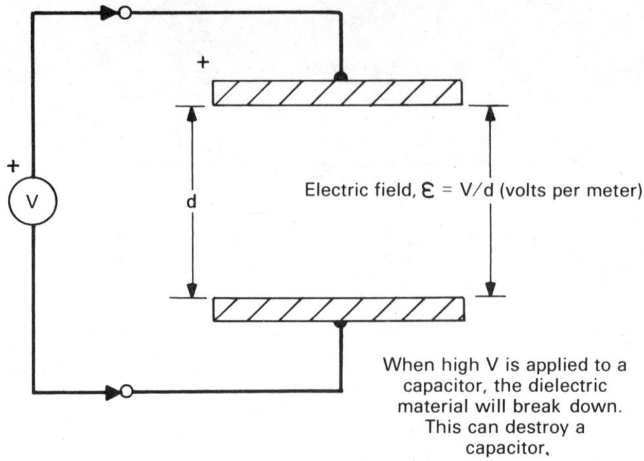

Figure 4.41 The voltage rating of a capacitor.

Sometimes a capacitor will even explode because of the internal gas pressure that is generated by the instantaneous decomposition of the dielectric material.

An air dielectric capacitor can momentarily break down, and if the voltage across the capacitor then returns to a much lower value, the spark will extinguish and the capacitor will survive. Oil, used as a dielectric material (an "oil-filled capacitor"), can also undergo momentary breakdowns without permanent destruction.

Lightning and the big capacitor in the sky

A common spectacular example of a capacitor in breakdown is a lightning discharge. This results from charge that the wind picks up from the surface of the earth and moves up onto cloud layers high in the sky, as shown in Figure 4.42.

The charge that is sitting in a storm cloud sets up an electric field between this cloud and the earth, as shown in Figure 4.43. It is this electric field that is trying to return that charge back to the earth where it came from. The presence of this electric field can be felt on a stormy, lightning-type day. Hair will stand up on your arms and perhaps even on your head because you are walking through the air dielectric layer of a large capacitor, a big capacitor in the sky.

When the electric field finally gets intense enough, this capacitor breaks down, and the flow or the return of this charge back to the earth, Figure 4.44, constitutes the large current surge that is the visible lightning spark. Thunder results from the noise the air makes when it abruptly flows back into the "hole" or tunnel that was created in the

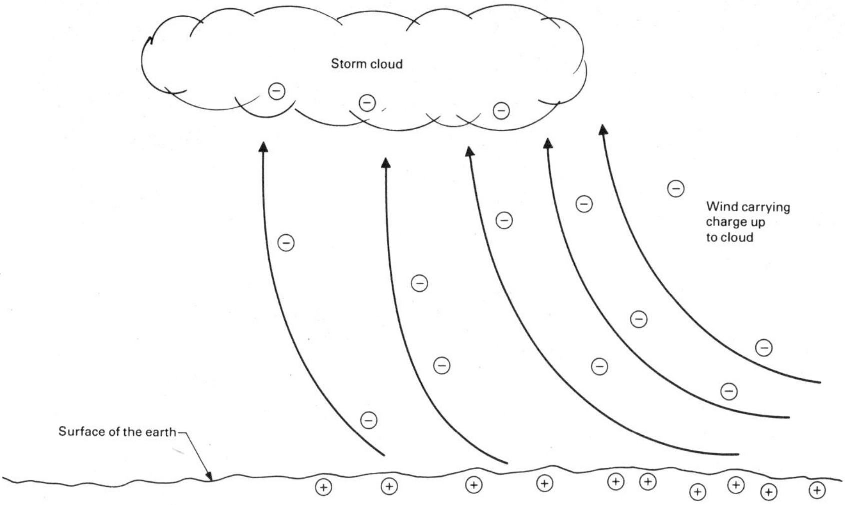

Figure 4.42 Charging the big capacitor in the sky.

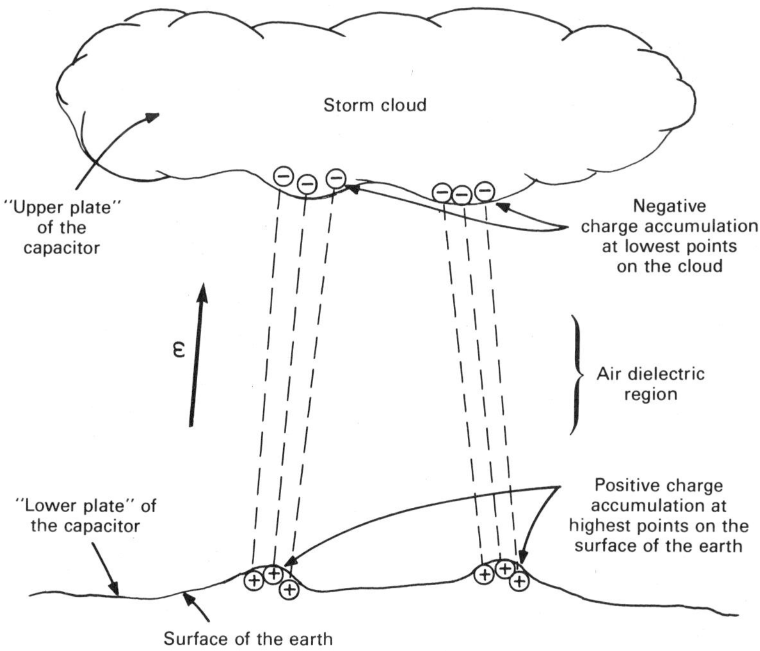

Figure 4.43 The electric field that results and attempts to return the charge to the earth.

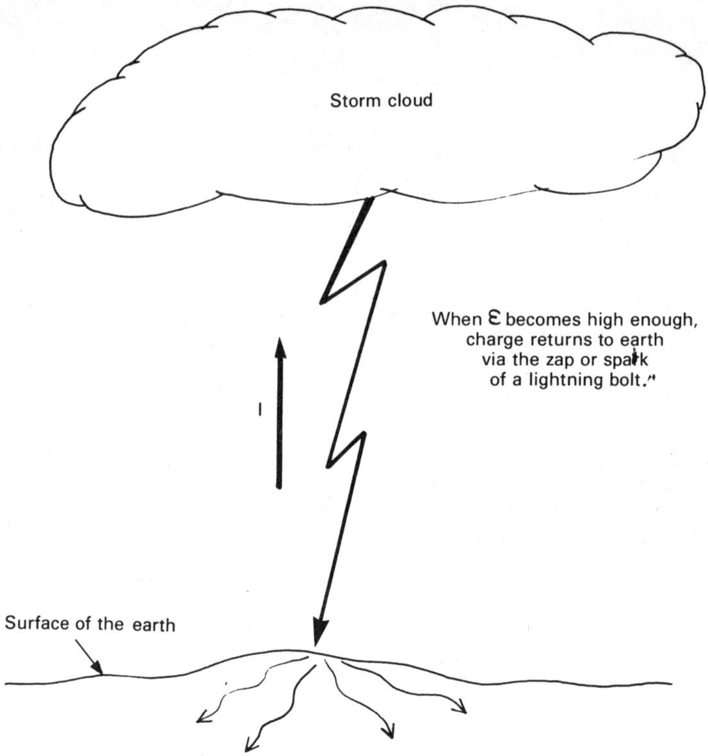

Figure 4.44 The lightning bolt ruptures the dielectric as the charge quickly returns to earth.

spark path. Storm clouds can also discharge from one cloud layer to another. This is called "brush lightning."

In the tropical areas, storm clouds exist at higher elevations and therefore more charge can be accumulated on these cloud layers before the critical electric field is reached and the lightning discharge takes place. As a result, the surge currents (the peak currents that flow) have been estimated at 400,000 A, an extremely large value of current. Voltages have been estimated at 300,000,000 V. This is why it's very dangerous to be outside walking around in a lightning storm (especially if you happen to have a pointed head). Your body represents a sharp projectile on the surface of the earth and therefore may be the shortest path for an electrical discharge. A lightning rod, a long metal rod that projects higher than the top of a house, uses this "shorter path" concept to protect the house. If a lightning discharge should attempt to take place in the vicinity of the protected house, it will be directed down the good conductor of this taller metallic rod and will not rip up the house.

A capacitor made from common materials

A capacitor can be made out of common materials that exist in a home. This is shown in Figure 4.45, in which two sheets of aluminum foil about 2 feet (ft) long and 1 ft wide are used. One of these sheets is put down on a table and is then covered with a dielectric layer (either a thin clear plastic food-storage wrap or waxed paper). The second sheet of aluminum foil is placed on top of this dielectric layer. This second aluminum foil is positioned so that it does not contact the first sheet of aluminum and also so that the long edges of each aluminum sheet extend out without lying over the other sheet (one across the top and the other across the bottom). On top of this second aluminum sheet, a second layer of either the plastic wrap or waxed paper is placed. This "sandwich" is folded up (as shown in Figure 4.46*a*), or it can be rolled up. The idea is to keep the two aluminum sheets from touching each other; this would short out (and thereby destroy) the capacitor.

One sheet of the aluminum has been shifted up a little bit with respect to the other sheet so that, when folded up, a short part of one aluminum sheet sticks out one end and the other sheet completely covers the rest of this sheet and a contact is made to both ends. These two ends become the contacts to the capacitor.

In Chapter 6 we will measure the value of this capacitor and will find that it is approximately 0.04 μF. This provides an appreciation for what it takes to make a capacitor of this value. A 1-μF capacitor would require a sheet of aluminum foil that is 25 times larger. Try to imagine this same technique being used to build a 1-F capacitor. This would have to be a million times larger than the 1-μF capacitor or 25 million times larger than the 0.04-μF capacitor!

The breakdown voltage of this homemade capacitor will be fairly large because the breakdown of plastic is a few hundred volts per mil (0.001 in). So, even with 0.5-mil plastic wrap, the breakdown voltage

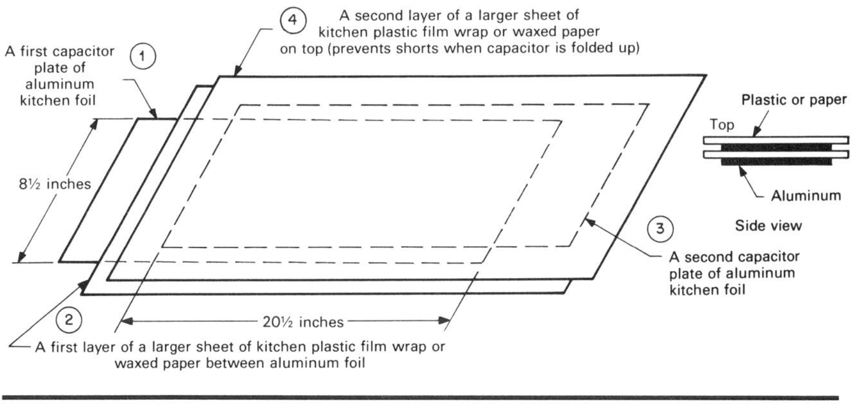

Figure 4.45 Making a plastic film or a waxed paper fixed capacitor.

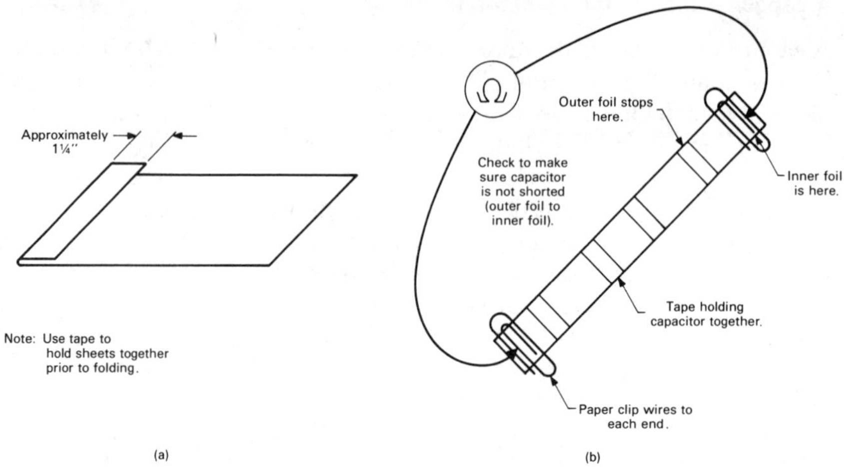

Figure 4.46 Folding up your capacitor. (*a*) Starting the foldup and (*b*) the finished capacitor.

will be in excess of 100 V. With 3-mil-thick waxed paper (that has a breakdown rating of approximately 200 to 300 V per mil), the breakdown voltage will be 600 to 900 V. These are large breakdown voltages because of the relatively thick dielectric layers. The benefit of a thinner dielectric layer is that breakdown voltage can be directly exchanged for a larger capacitance value.

Charge and a hydraulic analogy

A capacitor is very similar in its functioning to a water tank that accumulates and stores water. Water can be added to the tank at random times, as shown in Figure 4.47*a*. The tank is a reservoir. A spigot at the bottom of the tank can draw off a continuous stream, a "dc flow" of water, even though water is added to the tank in random quantities and at random times.

A capacitor, in an analogous electronic circuit, is shown in Figure 4.47*b*. "Charge packets" can be added at random time intervals and these charges will accumulate on the capacitor. The capacitor is storing this charge. A steady flow of charge (a dc current) can be extracted from the capacitor. This current can be passed through a resistive load, as shown in the figure. A large current flow (like 1 A) represents 1 C of charge extracted from the capacitor each second (1 A = 1 C of charge flow per second). Such a large current would rapidly discharge the capacitor.

A hydraulic circuit analogy is shown in Figure 4.48. Figure 4.48*a* is a hydraulic circuit that uses an elevated tank of water. The height of

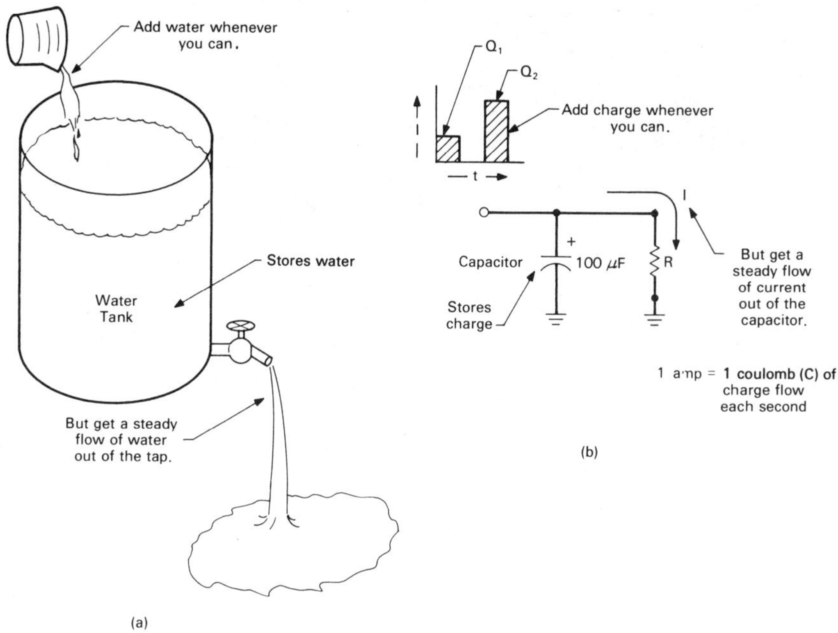

Figure 4.47 Capacitors are like a water tank. (*a*) The water tank and (*b*) the capacitor.

this tank provides the *pressure* that forces the water through the pipes and the plumbing fixtures of the hydraulic circuit. An ON/OFF valve is shown at the bottom of the tank. A small-diameter pipe (coming down from the tank) limits the flow of water. This small pipe is analagous to a large value of resistance in an electronic circuit.

A person is shown sitting in a bathtub at the bottom of Figure 4.48*a*. This is the purpose of the water flow and is analogous to a load in an electric circuit. Notice that water is drained at the bottom of the bathtub. A pump is used to lift this water back up to the tank by way of a separate pipe. This provides a closed circuit for the water flow. A relatively small-diameter return pipe is again analogous to a large resistance.

An electric circuit analogy of this hydraulic system is shown in Figure 4.48*b*. The load in this case is a light bulb. Current flow will light the lamp. A battery replaces the water pump and "pumps electrons up," and the voltage created by these displaced electrons causes them to flow through the circuit so that they can return to the battery. As in the hydraulic circuit (which had a continuous flow of water), this electric circuit has a continuous flow of electrons (current).

A resistor (R_1) tied between the battery and the capacitor is analogous to the small-diameter return pipe between the pump and the top

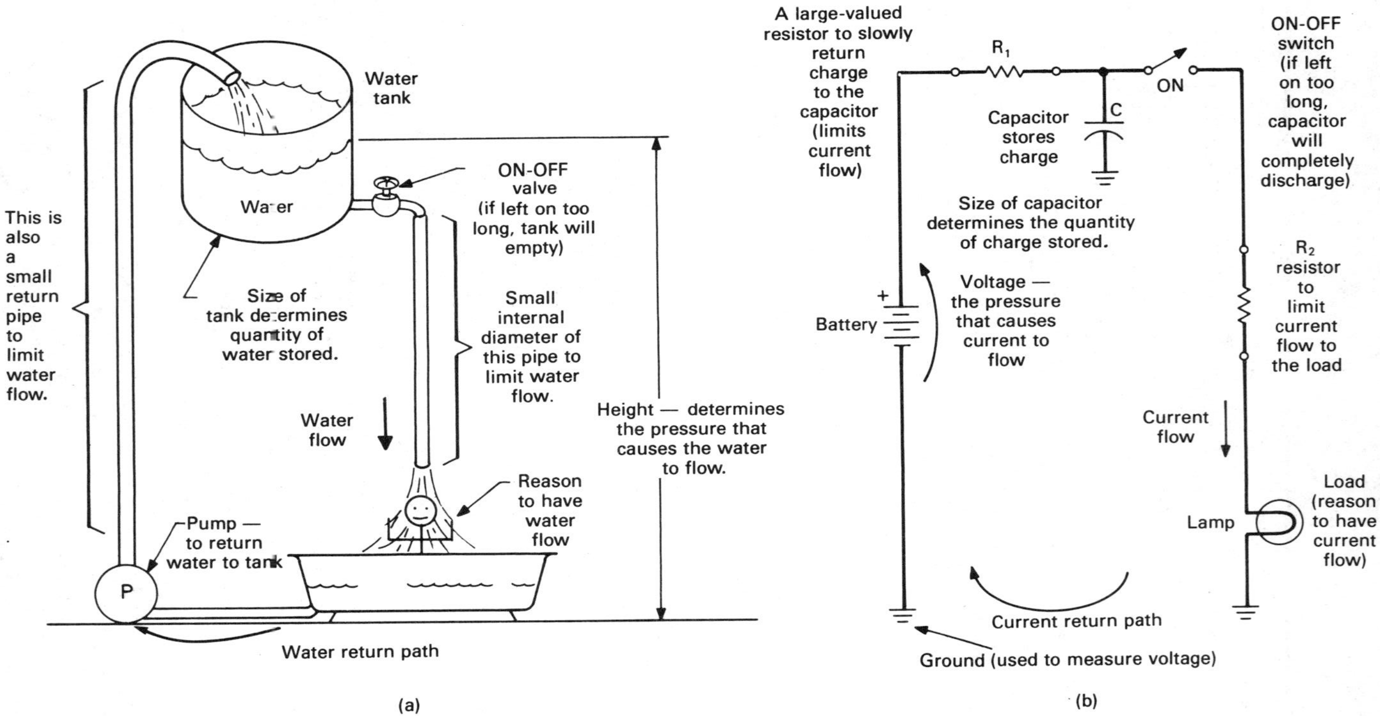

Figure 4.48 The analogy between a hydraulic circuit and an electrical circuit. (*a*) A hydraulic circuit and (*b*) an electric circuit.

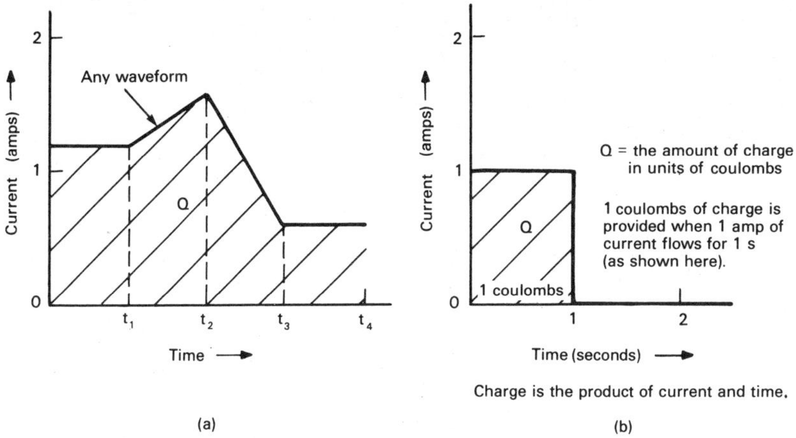

Figure 4.49 The relationships between current and charge. (*a*) A varying current and the resulting charge (*b*) One coulomb of charge.

of the water tank in the hydraulic analogy. This resistor is limiting the rate at which charge is returned to the capacitor. The resistor (R_2) tied between the battery and the lamp is analogous to the small-diameter exit pipe that limited water flow to the bathtub.

Capacitors accumulate and store "charge." Charge is sometimes a hard thing to think about. Voltage, current, and resistance are easier concepts, but what is charge? A current flow for a time interval constitutes a movement of a quantity of charge.

A couple of current waveforms are shown in Figure 4.49. Many different types of current waveforms exist. In the example shown in Figure 4.49*a,* the current is considered to start flowing at time 0 and is a constant current of 1.2 A up to time t_1. At t_1, the current linearly increases and has a value of 1.6 A at time t_2 and then linearly decreases to 0.6 A (600 mA) at time t_3, where it levels off and holds constant.

Charge is the area under a current curve, as shown in Figure 4.49*a.* Charge depends on how much current flows and also the time duration during which the current flows. Charge is the product of current and time. The units of charge are coulombs. One coulomb of charge exists when 1 A of current flows for 1 second (s). Figure 4.49*b* shows the example of a constant current flow of 1 A of current from time 0 to 1 s. This is the movement of 1 C of charge.

Capacitors as charge reservoirs

Capacitors can be used as charge reservoirs as shown in Figure 4.50. The Select switch (SW_1) of the lab kit is used to first charge (while in

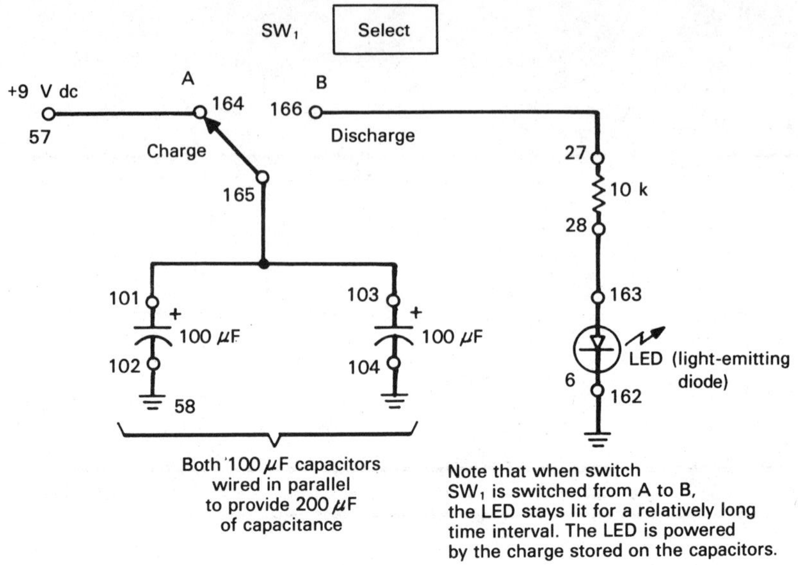

Figure 4.50 A capacitor used as a charge reservoir.

position *A*) both of the 100-μF capacitors to approximately 9 V. These capacitors are wired in parallel to provide a total capacitance of 200 μF.

When SW_1 is switched over to the *B* position, these capacitors will discharge through a 10-kΩ resistor and a light-emitting diode (diode No. 6 that is available via terminals 162 and 163). This LED is a way to provide a visual indication of the flow of current. The brighter the illumination, the higher the current flow.

Keep your eye on the LED as switch SW_1 is moved to the *B* position. The LED comes on most brilliantly at the start and then slowly tapers off in light intensity, but it does stay lit. Light is provided by this diode for a relatively long time interval. This light is provided by the charge that was stored on the capacitors. These capacitors are acting like a battery (they provide a source of voltage and current) but only for a relatively short interval of time. Capacitors cannot act like a battery over long time intervals unless only a very small amount of current is extracted. Capacitors store charge. They first accumulate this charge from a voltage source and then, at a later time, they can deliver this charge to an electronic circuit.

In the application of integrated circuits, a capacitor is usually needed from the power-supply voltage pin of the IC to ground, as shown in Figure 4.51. This capacitor typically has been called a "reservoir capacitor" because its function is to hold charge in reserve. If the IC has an instantaneous requirement for a large quantity of charge (a short burst

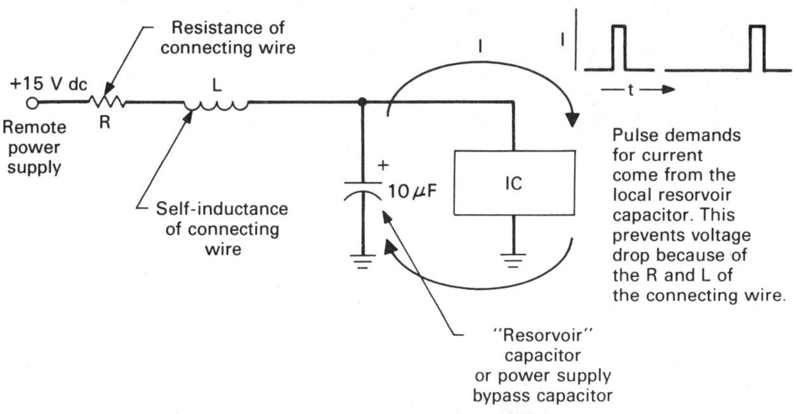

Figure 4.51 The capacitor reservoir for an integrated circuit.

of current flow), this charge will be supplied by the local reservoir capacitor. This prevents the relatively large voltage loss that could otherwise result if a large transient current were drawn through the long wires that connect to a remotely located power supply.

There are also circuits that will charge a capacitor and then later discharge it. This is a way to generate a very high peak current from a relatively small battery. An example of this is in the electronic strobes that are used with cameras. The voltage of a relatively small battery inside the strobe unit is greatly increased in magnitude and is then stored across a capacitor. When a flash is needed, the energy stored by this capacitor is abruptly discharged to cause a short-duration large current flow through a special light-generating tube. This creates the bright flash of light that is needed to expose the film.

Charging a capacitor

Figure 4.52 shows a test setup that is used to charge a capacitor. All of the cells in the battery section have been placed in series to obtain the highest voltage possible, 9 V. Also, the largest resistor available in the lab kit is used, 470 kΩ.

The capacitor uses the Key switch (with a 100-Ω series resistor) to discharge this capacitor (to bring the voltage across this capacitor back to approximately zero). When the Key is opened, the capacitor again will charge up. The dc voltmeter can be connected across the terminals of the capacitor to allow this charging to be observed.

When the Key is first opened, the voltage across the capacitor will increase relatively rapidly and then will slowly taper off and increase only very slowly as time goes on. This resistor-capacitor (*RC*) network allows the capacitor to be slowly charged from the 9-V dc voltage source.

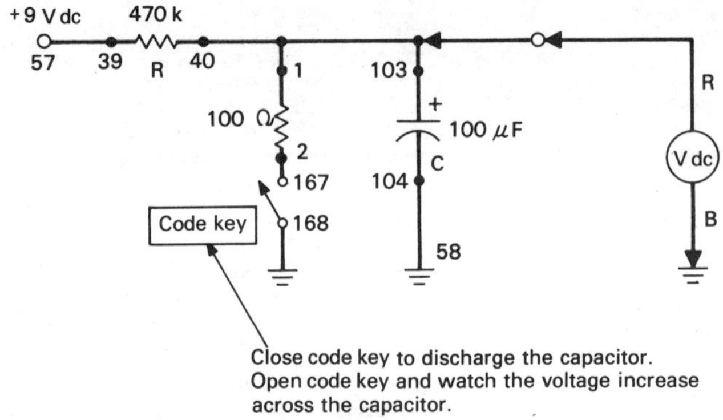

Figure 4.52 The test setup to charge a capacitor.

The voltage across the capacitor in this *RC* network follows a very precise exponential curve as shown in Figure 4.53*a*. The time units on this graph are given as multiples of the *RC* product (it may seem strange that when ohms are multiplied by farads, the units become seconds, but this is the case). This *RC* product is often called the "time constant" of an *RC* network. In one time constant the voltage across the capacitor will increase from 0 to about 63 percent of the final voltage. After two time constants, the voltage will be up to 86 percent. After three time constants, to 95 percent, and after five time constants, to 99 percent of the final voltage.

The capacitor will have acquired 50 percent of the final voltage after a time interval of approximately 0.7 *RC*. This can be used as a way to measure the value of a capacitor. If an *R* of 470 kΩ is used, the equation in Figure 3.52*b* shows that the capacitance value (in microfarads) is simply equal to 3 times the time (in seconds) that it takes for the capacitor to charge from 0 V to half of the final voltage. As an example, if one of the 100-μF capacitors is used, the time to charge from 0 V to half of the final voltage (4.5 V using a 9-V supply) would be 33 s. With a time of 33 s, the capacitance value can be calculated to be 99 μF.

It may seem that we could use this charging time as a general way to measure all capacitance values. We run into some problems with small values of capacitance because then we find that our 470-k resistor is not large enough. A small capacitor will charge up much too rapidly for us to accurately measure the time interval using our wrist watch or by simply counting seconds. Also, when resistance values get very high and capacitance values get very low, there are measurement problems because of stray voltage pickup. This stray pickup usually comes from the 60-cycle power lines, because they induce ac voltages into our measurement circuits. Therefore, we would have difficulty mea-

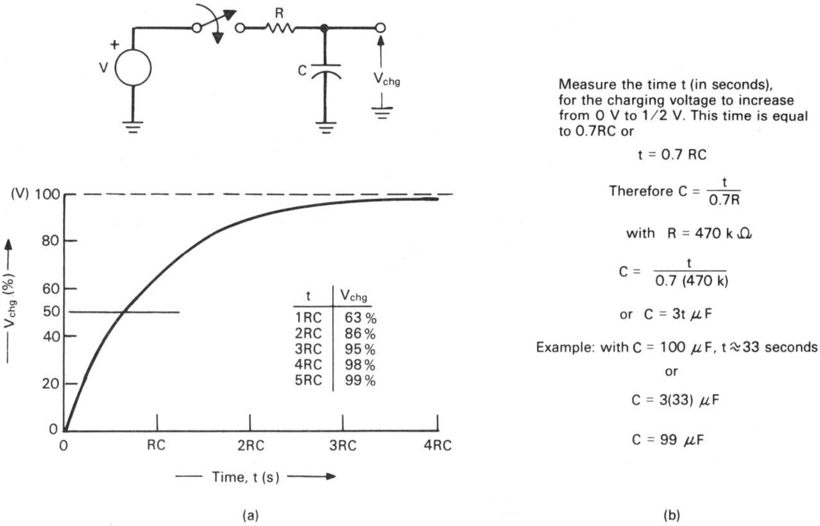

Figure 4.53 Measuring the charging time to determine capacitance value. (*a*) The charging waveform and (*b*) determining the value of capacitance.

suring small-valued capacitors using this technique and would have to carefully build the measurement circuitry inside of an enclosed, metal, shielding box and keep everything well protected from stray pick-up from the power lines.

Another alternative is to use an oscilloscope to measure the shorter time intervals. Oscilloscopes allow us to look at voltage waveforms on a screen (much like on a TV set). They provide the capability to relatively accurately measure times that are very, very small. In fact, modern oscilloscopes can measure time intervals smaller than a nanosecond; that's 10^{-9} s, a thousandth part of a millionth part of 1 s. Without an oscilloscope and a shielded box, we would have difficulty using this time-constant technique as a way to measure small capacitor values.

Stray capacitance delays voltage changes

Capacitors tend to resist voltage changes. This creates a major problem in the design of high-speed digital circuits. Even when capacitors are not specifically used, there is undesired "stray," or "parasitic," capac-

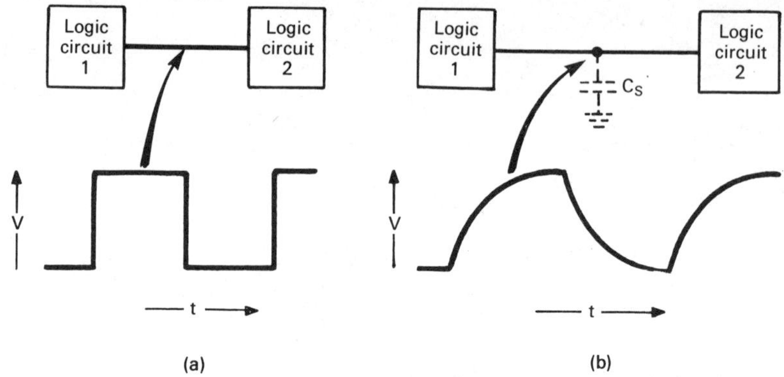

Figure 4.54 Effects of stray capacitance. (*a*) Ideal—no stray capacitance and (*b*) actual—with a very large amount of stray capacitance.

itance associated with the circuit interconnect wiring and the transistors and resistors that are used to build circuits. The effects of this stray capacitive loading on a logic circuit are shown in Figure 4.54. In Figure 4.54*a*, with no stray capacitance (an idealization), the logic voltage changes are very abrupt. With stray capacitance present, Figure 4.54*b*, the voltages cannot change as rapidly. This more slowly changing logic voltage waveform undesirably reduces the speed of the propagation of logic signals (voltage changes) through a digital system and therefore slows down the processing speed. Undesired stray capacitors are also a problem in the design of linear circuits because this capacitance delays the propagation of the high-frequency signals as they pass through a cascade of amplifier stages.

Capacitors sharing charge

A water analogy of capacitors sharing charge, Figure 4.55, shows how water is shared between two equal-sized glasses. Figure 4.55*a* shows one glass initially filled with water and the second glass initially empty. When the water is shared between the two glasses, both glasses end up half full.

In Figure 4.55*b*, a large glass and a small glass are used. At the start, the large glass is full and the small glass is empty. The water is shared by bringing the level in both glasses up to the same height. The water level of the large glass will not change much because the small glass does not hold much water.

In Figure 4.55*c*, the small glass is initially full of water and the large glass is initially empty. When the water is shared, the small glass will lose a considerable amount of the original water level because the large glass takes a lot of water to increase its water level.

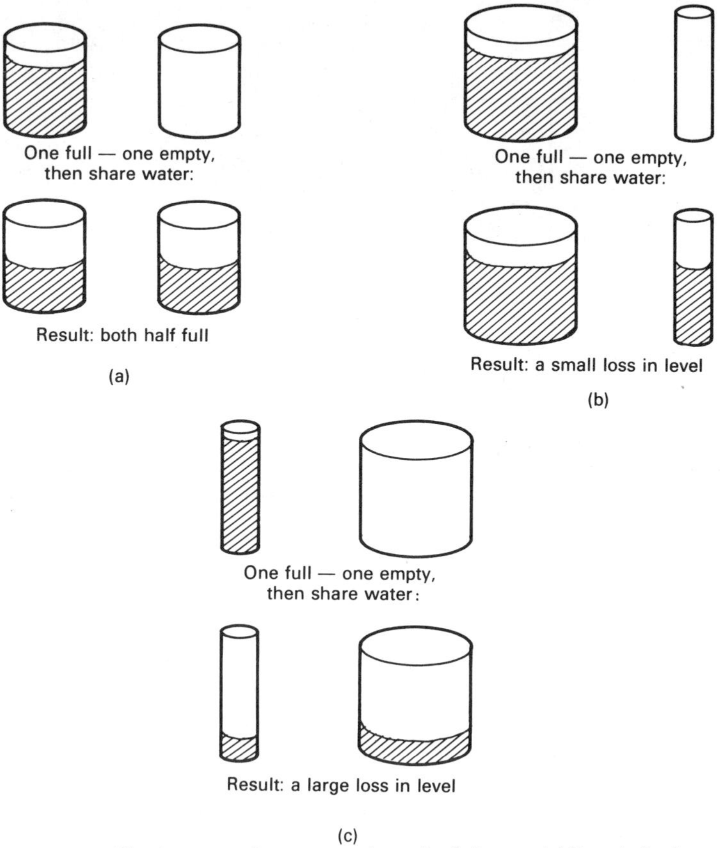

Figure 4.55 Sharing water between various-sized glasses. (*a*) Equal-sized glasses sharing, (*b*) a large glass sharing with a small glass, and (*c*) a small glass sharing with a large glass.

This water-glass analogy holds up identically with capacitors when they share charge. The analogy of the water is charge: glasses store water and capacitors store charge.

In Figure 4.56, two equal-valued capacitors are used. Both are 100 μF. With the Select switch (SW_1) in the *A* position, one of the 100-μF capacitors is fully charged to the 1.5-V level. While this capacitor is charging, close switch SW_2 (the Key switch) to discharge the second capacitor to make sure that it is initially completely "empty."

Notice that whenever a capacitor is discharged, a 100-Ω resistor is placed in the discharge path. The purpose of this small-valued resistor is to provide a limit on the maximum discharge current that is allowed to flow. Without this resistor, the contacts of the switch could be damaged because of the large initial surge current. It is always advisable to place some resistance in series with switches that are used to discharge capacitors, provided that no detrimental effects in the overall performance of the circuit result.

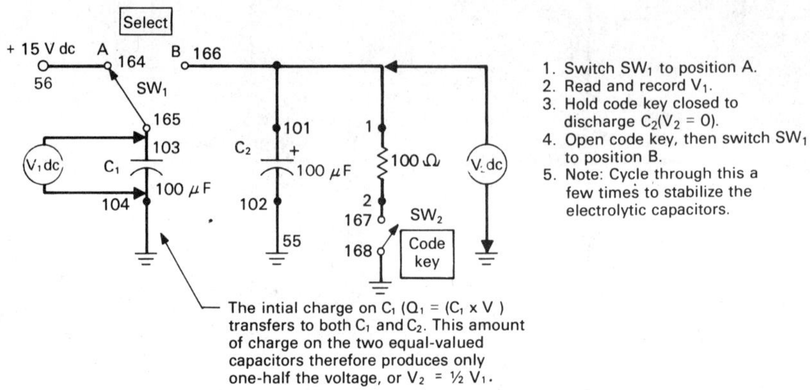

Figure 4.56 Sharing charge between equal-valued capacitors. (*a*) Test setup and (*b*) procedure.

Close the Key discharge switch (SW_2) and watch the voltage across capacitor C_2 go to zero. Open the Key switch, then slide the Select switch (SW_1) over to the *B* position. Notice that equilibrium will be immediately reached. Both capacitors will have a voltage of about half of the starting voltage (something close to 0.75 V). This experiment consists of two equal-valued capacitors sharing charge and is analogous to two equal-sized glasses sharing water.

A second example, shown in Figure 4.57, is a large-valued capacitor sharing charge with a small-valued capacitor. Notice that the large-valued capacitor is 100 μF and the small-valued capacitor is only one-

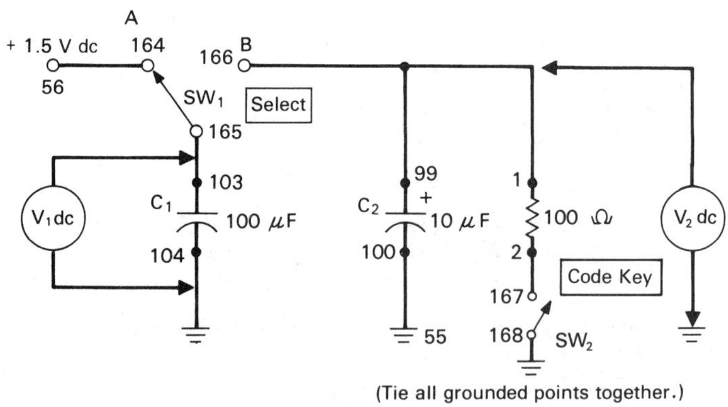

Follow same switching sequence as on previous figure. Note that not much additional capacitance is added now (C_2 is only 10 μF and C_1 is 100 μF), so V_2 is only slightly less than V_1.

Figure 4.57 Sharing charge from a large-valued capacitor to a small-valued capacitor.

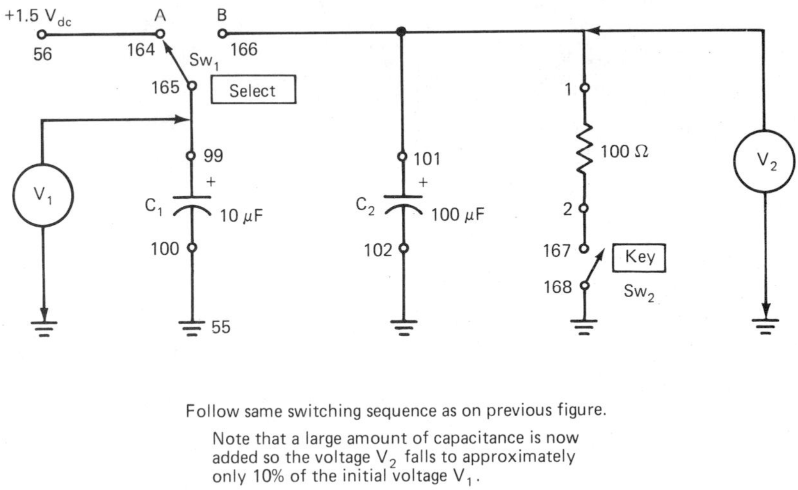

Figure 4.58 Sharing charge from a small-valued capacitor to a large-valued capacitor.

tenth as large (10 μF). In this case, when the charge is shared, the voltage across the large capacitor will not fall very much because only a small charge is supplied to the small-valued capacitor.

To complete the analogy, the capacitors are interchanged and the 10-μF capacitor is charged. We are concerned about the final voltage that results when this charge is shared with the 100-μF capacitor, as shown in Figure 4.58. Notice that in this case most of the initial voltage is lost.

Capacitors provide voltage memory

A water tank can be filled up, and if left alone, it will "remember" the amount of water that it contains. In an analogous way, capacitors can "remember" voltages. Capacitors are commonly used to store voltage levels. In dynamic random access memory (RAM) ICs, the digital memory is obtained by using many small-valued capacitors to store voltages.

As an example of voltage memory, Figure 4.59 shows that the 100-μF capacitor can be charged when the Select switch (SW_1) is in the *A* position. While switch SW_1 is still in the *A* position, the Key switch (SW_2) can be closed and the voltmeter will indicate that the full 9 V is across this capacitor. Then the Key switch is opened, and the Select switch (SW_1) is moved to position *B*. The capacitor is now sitting with nothing connected to it, so it will remember or retain this initial voltage.

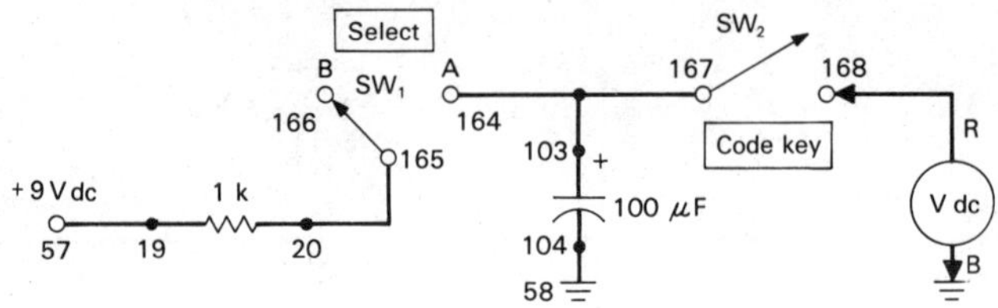

Procedure:

1. Charge the capacitor to the maximum voltage (hold code key switch closed and read on meter) by closing SW_1 (to position A).
2. Open the code key switch, then open SW_1 (to position B).
3. Momentarily close the code key switch long enough to read the voltage across the capacitor—then open the code key switch again.
4. Wait many seconds.
5. Again, repeat step no. 3 and note how much voltage remains across the capacitor.
6. Again, wait; and again read the voltage. Notice how long the voltage is maintained by this 100-μF capacitor.

Figure 4.59 Capacitors provide voltage memory.

The voltmeter can be connected to the capacitor via the Key switch to momentarily read the voltage. The reason for using this momentary connection of the voltmeter to the capacitor is because the voltmeter does draw a small amount of current and therefore slightly discharges the capacitor. The voltmeter has an input resistance (between the red and black measuring leads) of about 10 MΩ. The Key switch keeps this 10-MΩ load away from the capacitor: it connects the voltmeter for a short time interval, only long enough to obtain a reading.

The capacitor is charged and the initial voltage is measured. Then, after each of a sequence of long time intervals, the voltage is quickly measured. Notice that not much loss in voltage takes place between these measuring intervals.

Capacitor specifications often indicate this ability to retain voltage for long time intervals by a parameter that is called the "self-time constant." A self-time constant (RC) of a capacitor represents the product of the capacitance value (C) and an equivalent external resistor (R) that would draw a current equal to the internal dc leakage current of the capacitor. Self-time constants of the best storage capacitors (Teflon capacitors) can be 50 years! This means that after 50 years, a capacitor with a Teflon dielectric, left alone with nothing tied to it, will lose 63 percent of its initial voltage. The self-time constant is usually highly temperature dependent. If the temperature is elevated, the self-time constant will be reduced because leakage resistance typically decreases as temperature increases.

Capacitors can provide time delays

Capacitors are often used in electronic circuits to provide time delays. A circuit that demonstrates one way of doing this is shown in Figure

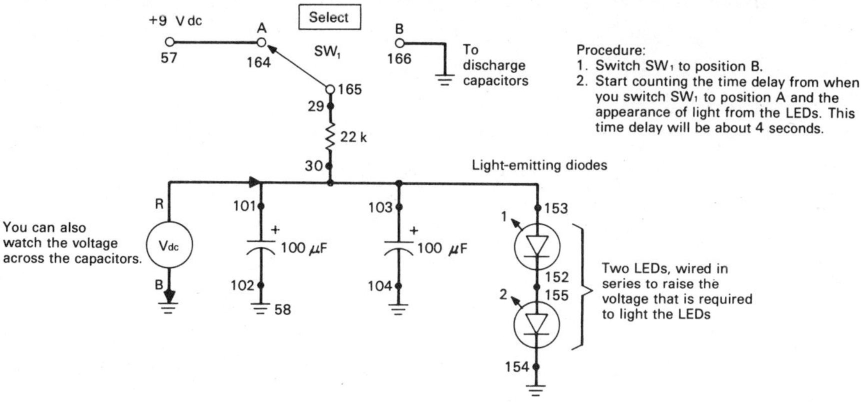

Figure 4.60 Capacitors causing a time delay. (*a*) Test setup and (*b*) procedure.

4.60. When the Select switch (SW_1) is moved to position *A*, the capacitor initiates a time delay before the LEDs light. Two LEDs are connected in series so that the LEDs will take twice the voltage of a single diode before these diodes turn on. In this example, to get a large-valued capacitance, the two 100-μF capacitors of the lab kit are connected in parallel.

Start with the Select switch (SW_1) in the *B* position to fully discharge the capacitors. After the capacitors are fully discharged, move the Select switch (SW_1) to the *A* position (which will apply 9 V to this *RC* network). The 22-kΩ resistor is creating a slow (exponential) buildup in the voltage across the capacitors. The turn-on voltage of the LEDs keeps them off until about 3 V has been accumulated across the capacitors. This creates the time delay. Change SW_1 from position *B* to position *A* and start counting time. Notice that about 4 s goes by before the LEDs turn on. This circuit has a 4-s time delay from the closure of the switch to the appearance of light out of the LEDs.

Capacitors block dc current

Good capacitors do not pass dc; dc current does not flow through a capacitor. Once a capacitor charges up, there will no longer be any current flow from a dc voltage source.

Electrolytic capacitors, on the other hand, often pass a dc leakage current. The oxide layer in an electrolytic capacitor is constantly being formed while the capacitor has a voltage across it. The forming of this oxide creates a small current flow. But even after this oxide layer is completely formed, a small-valued current will still flow. This small current flow can be determined *indirectly* by measuring the voltage that this current develops across a 100-kΩ current-sampling resistor.

Build the circuit shown in Figure 4.61 and then first move the select switch (SW_1) to position *B* to completely discharge the capacitor. When switch SW_1 is first moved to the *A* position, the voltage reading across the 100-k resistor will be very close to the full 9 V. This voltage results from the surge of current that is initially charging the capacitor. This voltage will decrease very slowly as the capacitor becomes charged. After about 3 min, this voltage can still be approximately 0.3 V (300 mV). This implies that there is dc current leaking through the capacitor that is about 3 μA (this calculation is shown on the figure). This is the initial oxide-reforming current flow of the electrolytic capacitor. After a few hours more, the voltage reading should drop to about 13 mV, indicating a current flow of only 130 nA.

The printed circuit board material in the lab kit is made out of cardboard, and so something like 10 nA of current flow can be measured across the surface of this board. This current has nothing to do with the electronic components. If the leakage currents of some of the smaller-valued capacitors are measured, there will be difficulties because of 60-cycle "hum" pickup from the power line wiring and also because the actual value of current that you (indirectly) measure will never be smaller than this board leakage value.

After charging this capacitor, move the Select switch (SW_1) to the *B* position, which will ground the positive side of the capacitor. The opposite side of the capacitor therefore will instantly drop to −9 V. These negative voltages can be measured with the voltmeter as the capacitor discharges. After a long time has elapsed, the capacitor will completely discharge and the voltmeter reading will be zero. Many circuits make use of a capacitor to provide a negative voltage in a circuit when only a positive power supply voltage is used.

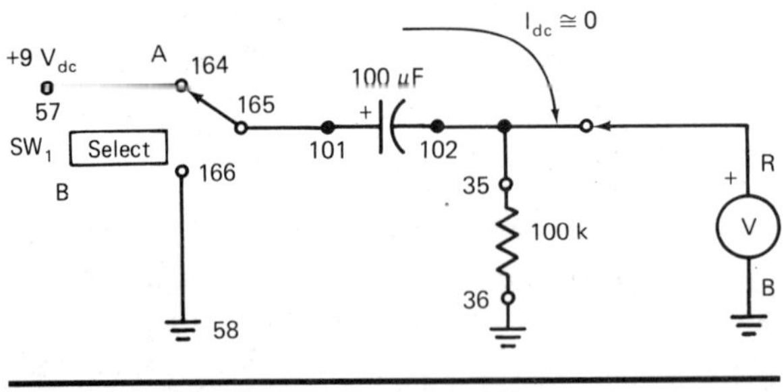

Figure 4.61 Electrolytic capacitors pass some dc current. (*a*) The test setup and (*b*) the procedure.

Piezoelectric Quartz Crystals: The Frequency References

When mechanical forces are applied to crystals that exhibit what is called the "piezoelectric effect," voltages are generated on opposite faces of the distorted crystal, as shown in Figure 4.62. There are many applications in electronics for these piezoelectric crystals, such as phonograph cartridges, earpieces (as in the lab kit), microphones, the new flintless and batteryless cigarette lighters, and the quartz crystals that are used to provide stable frequency references. The symbol, the physical mounting, and the use of a crystal are shown in Figure 4.63.

Quartz is the crystalline form of the oxide of silicon (SiO_2). It has very good dimensional stability during temperature changes. The

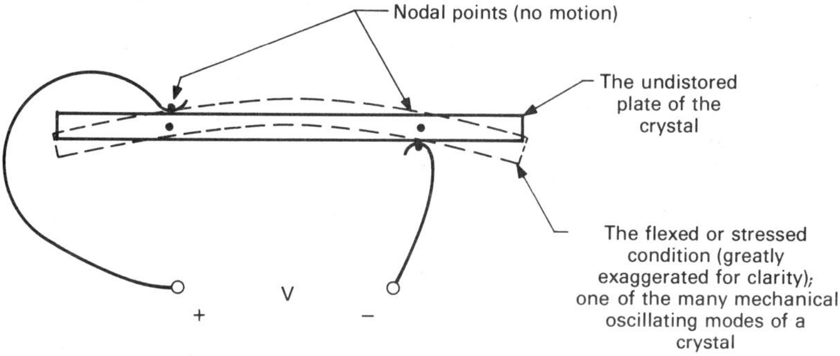

Figure 4.62 Voltages are generated by a stressed crystal.

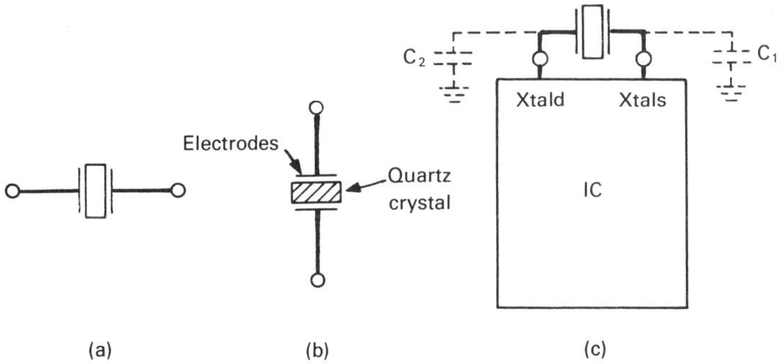

Figure 4.63 The piezoelectric quartz crystal. (*a*) Symbol, (*b*) actual arrangement, and (*c*) connection to an IC.

mechanical resonant frequency of a quartz crystal depends on its physical dimensions and is therefore stable over both time and temperature changes.

Many radio receivers (and also radio transmitters) make use of quartz crystals, especially when operating on fixed frequencies (channels) as in citizens' band radio. Also, the basic clock circuit in digital computers and digital communications circuits uses a quartz crystal because of the low initial frequency tolerance and the excellent temperature stability that is obtained.

Diodes: The Two-Terminal One-Way Devices

A diode is like a one-way street. Diodes allow current to flow in one direction but prevent current from flowing in the opposite direction. This action and the electronic symbol for a diode are shown in Figure 4.64. Notice that both an arrowhead and a bar make up the diode symbol. The arrowhead is the "anode," and the bar is the "cathode" of the diode. In less formal discussions, people might refer to the "arrow" and the "bar" because the words "anode" and "cathode" don't come to mind anymore. These words were common in the days of vacuum diodes. The direction of the arrowhead in a diode is the direction of the easy flow of current. Current will not flow in a direction opposite to this arrow.

This directionality of a diode is shown in Figure 4.65, in which an LED is again used as a visual indication of current flow. A 10-kΩ current-limiting resistor is placed in series with the LED. When the diode is connected as shown in Figure 4.65*a*, the current flow is blocked. If the connections are reversed (as in Figure 4.65*b*), current flows.

Some "logical" uses for diodes

Diode-logic circuits have been used in digital computers. Two diode-logic circuits are shown in Figure 4.66. Figure 4.66*a* shows an OR gate

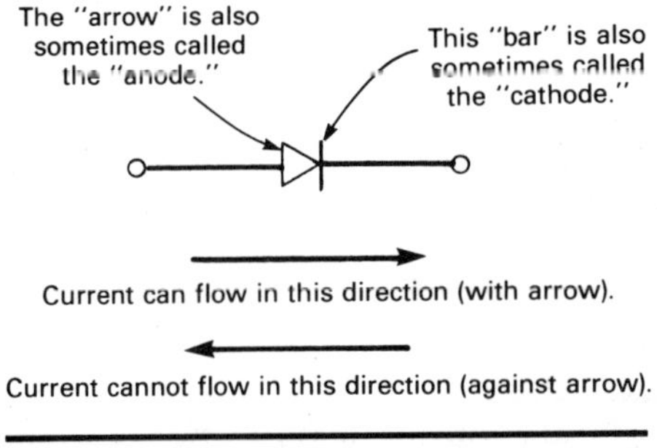

Figure 4.64 The symbol for a diode.

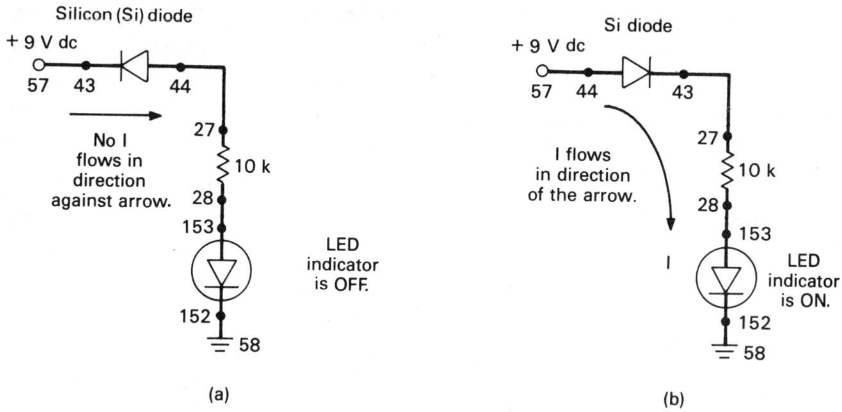

Figure 4.65 Diodes conduct current in only one direction. (*a*) Diode blocks current flow and (*b*) diode allows current flow.

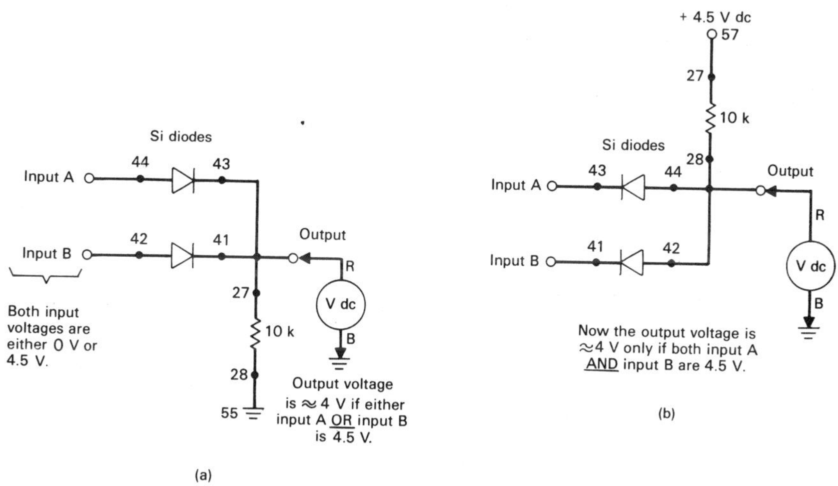

Figure 4.66 Diodes used in logic circuits. (*a*) The OR gate and (*b*) the AND gate.

with two inputs, *A* and *B*. Each of these digital input voltages can be either 0 or 4.5 V (4.5 V is the closest voltage that is available in the lab kit to the standard 5-V power-supply voltage). If the *B* input happens to be at 0 V and the *A* input is at 4.5 V, the diode tied to the *A* input will pass the 4.5 V over to the output. So, the output will be very close to 4 V (approximately 0.6 V is lost in going through a conducting silicon diode). The diode on the *B* input is reverse biased. The anode (arrow) is at ground, and the cathode (bar) is up at +4 V (so this diode is not conducting). This allows the *B* input to be at 0 V without disturbing the output voltage level. For this OR gate, if either input *A* *OR* input B (*OR both* of them) is at the high voltage of 4.5 V, the output will be in the high-voltage state. This is called the logical "OR function."

An AND logic gate is shown in Figure 4.66*b*. The 10-kΩ resistor now ties to the 4.5-V power supply and the diodes are reversed. If either input *A* or input *B* is at 0 V, the output will be pulled down to very nearly 0 V (it will actually be one diode drop above 0, or about 0.6 V). Both *A AND B* must be in the high-voltage state to get the output in the high-voltage state. When both *A AND B* are at 4.5 V, the output will also be at 4.5 V because both diodes are blocking. This allows the output voltage to rise to the power-supply voltage level.

A closer look at semiconductors

To understand the internal action of a diode, the earlier discussions of silicon will be reviewed. Silicon forms a crystal like carbon forms a diamond. A pure silicon crystal is not a good conductor. To be of use in electronics, other elements must be substituted for a few of the silicon atoms within an otherwise perfect crystal structure.

Figure 4.67 shows two useful elements, boron and phosphorus. The outermost shell of a boron atom has only three electrons (one less than the four that exist with silicon). Conversely, the outermost shell of a phosphorus atom has five electrons (one more than the four that the silicon atom has).

As shown in Figure 4.68, if a boron atom can somehow be substituted for a silicon atom, the three electrons that are provided by the boron atom will be shared by three of the adjacent silicon atoms. A problem occurs with the silicon atom shown to the right because this silicon atom has an electron to share but the boron atom does not. This represents a distortion in the crystal lattice because *a place exists for an electron.* If an electron can be found wandering around within the crystal structure, this place for an electron (called a "hole") can capture this extra electron and thereby satisfy the Si atom to the right.

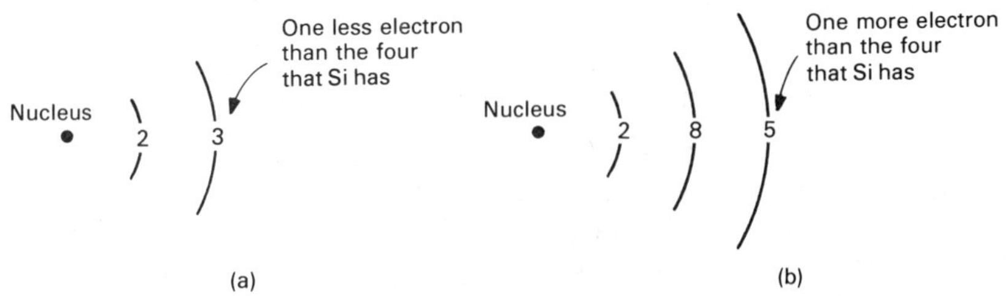

Figure 4.67 Dopants for silicon. (*a*) Boron (B), a P-type dopant and (*b*) phosphorus (P), an N-type dopant.

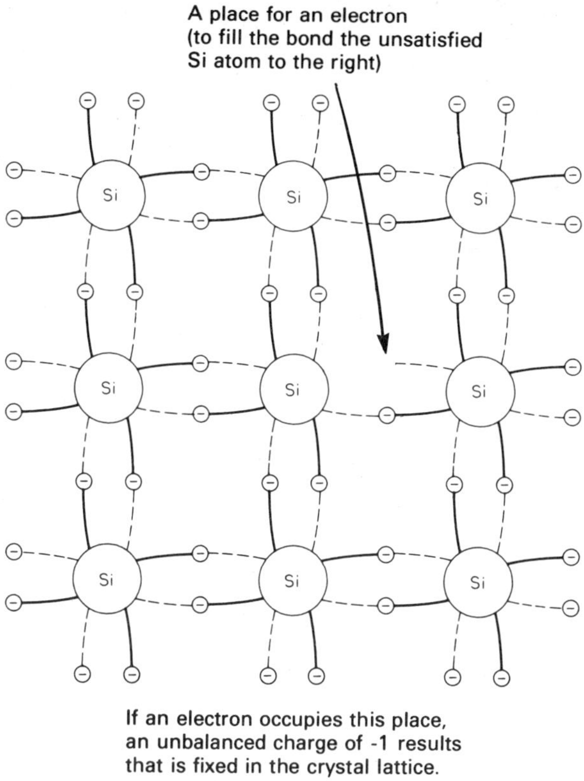

Figure 4.68 P-type doping of silicon.

This hole (this absence of an electron) is acting like a positive charge. Doping of silicon with boron is therefore said to produce "P-type" silicon (the "P" is for positive).

Figure 4.69 shows a phosphorus atom replacing one of the silicon atoms. There are five electrons in the outermost shell of a phosphorus atom. A phosphorus atom is therefore able to provide one electron in a covalent bond with each of the four neighboring silicon atoms and have one *spare* electron left over. This electron is shown very close to the phosphorus atom. In actuality, this spare electron can be located anywhere within a vicinity of a few thousand silicon atoms away because this electron is not very tightly held to the phosphorus atom. This extra electron is therefore easily dislodged, and it is then available to conduct current. Because of this extra electron and its negative (N) charge, phosphorus provides what is called "N-type doping."

By using controlled doping, both N-type and P-type regions can be formed within a silicon crystal. This is how PN diodes, NPN transistors, PNP transistors, N-type resistors, and also P-type resistors are

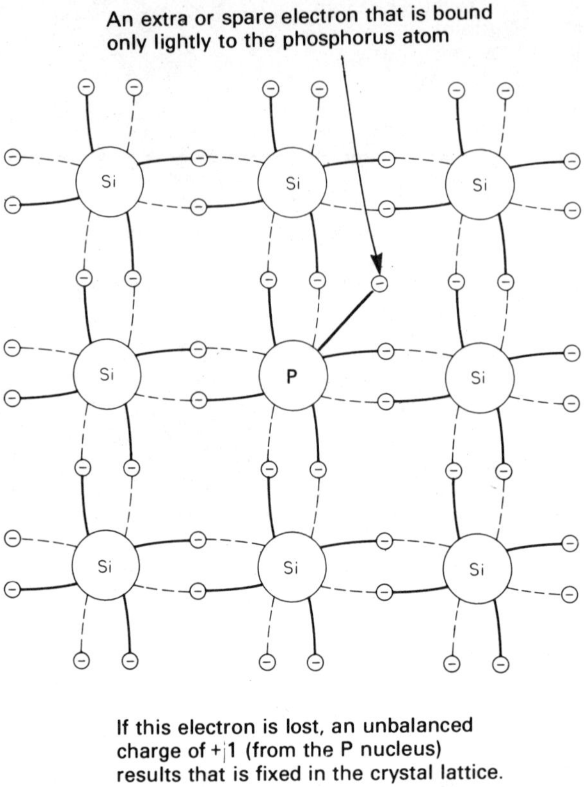

Figure 4.69 N-type doping of silicon.

made. All of these designations refer to the types of doping that were used in the silicon.

When silicon is doped, it is changed from a "semiconductor" to a "conductor" of electricity. A heavily doped silicon crystal is indicated as N^+ or P^+. Heavy doping provides low values of resistance. An N^- or a P^- designation represents light doping concentrations. Lightly doped silicon is not a very good conductor of electricity.

The PN diode

If two separate blocks of silicon, one doped P type and the other doped N type (as shown in Figure 4.70*a*), are fused together (Figure 4.70*b*), a PN diode is formed. The spare electrons that are available in the N-type side will immediately flow (by a process called a "diffusion current" flow which is caused by the concentration gradient of the electrons) across the PN junction between these two blocks of silicon and fill the holes in the P-type side. The question comes up, Will this diffusion of

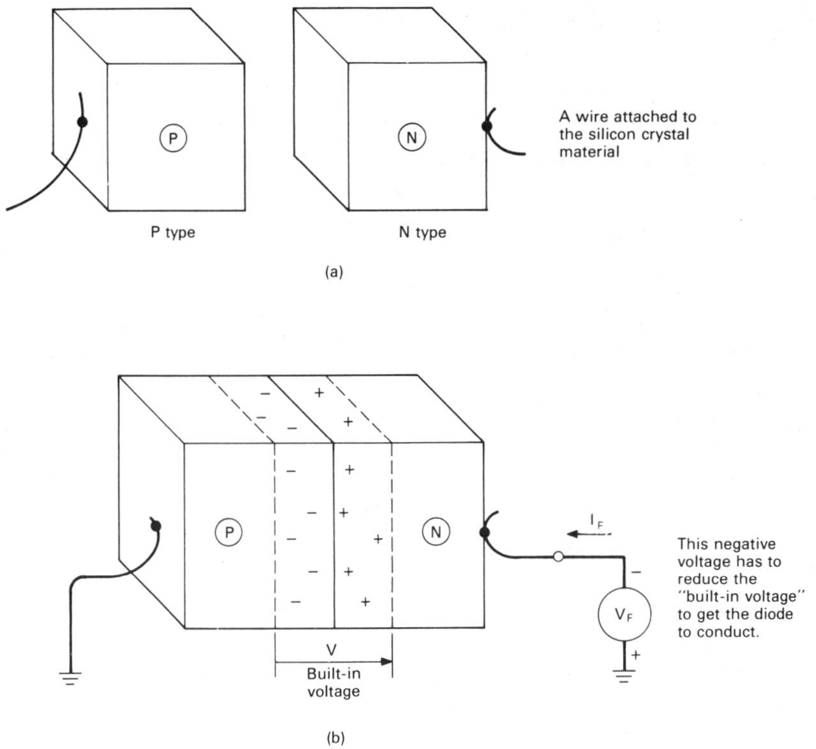

Figure 4.70 Making a PN diode. (*a*) Two separate crystals of doped silicon and (*b*) on contact, a PN diode is formed.

these electrons continue forever? What's to prevent the electrons from continuing to diffuse across the junction into the P side and then flowing out a wire attached to that side and returning back to the N side and continuing this electron flow forever in this circuit (a constant current flow with no batteries or other power source needed)? This would be a form of perpetual motion. There is no energy consumed, but could a steady flow of current exist that we could put to work powering a light?

The strange thing about this particular diffusion mechanism is that as each electron leaves the N-type side of the diode and then flows over into the P-type side, it leaves behind a positive charge within the nucleus of each of the dopant atoms that originally supplied the electron. When the N-type side loses those electrons, the phosphorus doping atoms (that are fixed in the crystal lattice) are unbalanced. The nuclei of these phosphorus atoms have a charge of +5. But the fifth electrons of these phosphorus atoms are now gone and are therefore no longer exactly balancing the total positive charge that exists within the nucleus of each phosphorus atom. These resulting positive fixed charges are therefore embedded within the silicon crystal lattice on the N-type side. So after a few electrons initially move away, an equal number of uncovered positive charges will result within the lattice on the N-type side of the diode, near the metalurgical junction. Similarly, when these electrons enter the P-type side, they will "fill up" those places that exist for electrons, the holes, and thereby also create an unbalance of fixed charge on this side of the diode.

What actually happens is that this separation of the fixed charges, which results from the initial mobile electron flow across the diode, establishes a voltage difference across the diode. This voltage difference establishes an electric field that is in a direction to return these mobile electrons to their original location. The presence of this built-in field also tends to prevent any further *diffusion* of the mobile electrons across the metallurgical junction. Therefore, a natural balancing action is set up in which, initially, mobile electrons diffuse across the junction because of the difference in the relative amounts of mobile electrons on each side of the diode. The unbalanced fixed charges that result within the crystal lattice then set up an electric field that continually is returning these mobile electrons back to the N-type side where they came from. This natural balancing of these two components of current sets up an equilibrium condition. This can sound very confusing because, for example, mobile electrons are always moving from left to right across the junction by a diffusion mechanism and are also returning from right to left (a "drift" current) by the action of the electric field. Note that the mobile charge that moves because of an electric field is a *drift* current, and the mobile charge that moves by diffusion is a *diffusion* current.

This balancing of the flow of these two types of current (drift and diffusion) within a PN diode is similar to the situation that exists when water is boiling in a covered pot on the stove. When "thermal equilibrium" is reached (this is when everything is at a stabilized temperature), the rate at which new steam is being created is the same as (or is balanced by) the rate at which existing steam is being condensed back into water on the underside of the lid and the interior walls of the pot.

If the lid were slightly removed, steam would escape and new steam would then be generated at a much higher rate than the rate of condensation back into water. The application of an external voltage that "forward biases" a PN diode is similar to *slightly removing the lid* from a pot of boiling water. A "reverse bias," then, is analogous to simply *more tightly holding the lid* on the pot.

Because of the built-in field that results from the separation of the fixed charges on each side of a diode, the polarity of the external voltage that is applied to a diode is very important. For example, if this polarity adds to the magnitude of the built-in field, it will increase the width of the region that has no mobile charge carriers, the "space-charge layer," and the diode will not conduct current (the lid is held more tightly on the pot). But, if the polarity is interchanged, the width of the space-charge layer and the strength of the built-in field would both become smaller. Therefore, because of this reduced built-in field, a continuous *diffusion* of charge carriers would allow current to flow through the diode (the lid is slightly removed from the pot). Remember that forward biasing voltages allow current to flow in a diode, but reverse biasing voltages do not allow current to flow. This is why diodes only conduct current in one direction.

In a real diode, usually one or the other of these mobile charge carriers will dominate. For example, an N^+-P^- diode would have more mobile electrons involved in providing for the forward-biased current flow than mobile holes, and a P^+-N^- diode would have more mobile holes involved than mobile electrons. But a diode with nearer the same doping concentrations in both the N-type and P-type sides will have both mobile carrier types involved. For simplicity, we will consider the situation that exists when the mobile electrons dominate.

But what happens if a reverse voltage is applied to a diode and then this reverse voltage is continually increased in magnitude? Will current ever flow? This is similar to the lightning example, in which charge was continually moved from the surface of the earth and placed on a cloud. Eventually, such a large voltage resulted between the cloud and the earth that the insulator (the air) that had been separating these two groupings of charge finally broke down and current did flow. Similarly, diodes will also enter breakdown if high reverse voltages are applied across them. Diodes in breakdown are also useful electronic components because the magnitude of this breakdown voltage can be used as a voltage reference (these are the "zener" diodes and the "avalanche" diodes).

The forward characteristics of diodes

The forward voltage that has to be applied to a diode to get it to conduct current can be measured as shown in Figure 4.71. Notice that two values of voltage (1.5 and 4.5 V) are used. A series resistor (R) limits the current flow to the diode and will be changed over the range from 470 kΩ to 470 Ω as larger forward current flows through the diode. The multimeter is set up to read dc volts and provides the data shown in Figure 4.71*b*.

The potentiometer labeled "Control" is adjusted to sequentially obtain the values of forward voltage (V_1) shown. First, adjust for $V_1 = 200$ mV, and then record the second voltage (V_2). Then adjust to obtain a

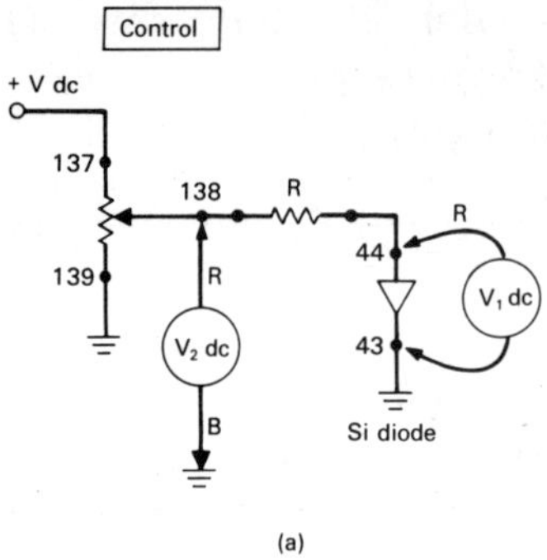

Diode voltage ↘ (V₁, mV) Diode current ↘ (I*)

+V dc	R	V_1 mV	V_2	$(V_2\text{-}V_1)$	I*
1.5 V	470 k	200	209 mV	9 mV	0.03 mA
		300	314	14	0.03
		400	460	60	0.128
		500	1.460 V	960	2
4.5 V	47 k	600	2.77	2.17 V	46
	4.7 k	700	3.73	3.03	645
	470 Ω	800	3.71.	2.91	6.19 mA

For highest accuracy, use actual measured values of these R's to calculate I.

$$*I = \frac{(V_2\text{-}V_1)}{R} = \frac{V_R}{R}$$

(b)

Figure 4.71 Obtaining data on the forward characteristics of a silicon diode. (*a*) Test setup and (*b*) tabulating the data.

forward voltage of 300 mV, and so on. V_1 steps of 100 mV are used over the range from 200 to 800 mV. The current flow (*I*) in the diode is calculated by differencing the two voltages (V_2–V_1) and then dividing this difference voltage (the voltage across the resistor *R*) by the value of the resistor *R*. This is an *indirect* way to measure the current *I*.

Looking at these data in a tabular form does not provide much of a "feel" for what is going on. It is better if these data are plotted so that relationships can be seen more clearly. Therefore the previously tabulated data on the silicon diode are plotted in Figure 4.72. Notice that a silicon diode requires a forward voltage of approximately 580 mV before any significant current flows, but when the diode does start to conduct, it has a fast turn-on characteristic. Substitute a germanium diode (also available in the lab kit) in this test setup and make a few measurements to see that the forward voltage of a germanium diode is less than that of a silicon diode.

The effects of ambient temperature on the forward voltage characteristics of a diode can be demonstrated with the circuit shown in Figure 4.73. A silicon diode is used and is biased with a 10-kΩ resistor from a 9-V power supply. The value of the forward voltage of this diode at room temperature is first recorded. Then the hair dryer is turned on and held about 2 or 3 in from the silicon diode. Notice that the forward voltage falls as the temperature of the diode increases. A good rule of thumb is that the diode forward voltage falls at the rate of about 2 mV for every degree Celsius (centigrade) increase in the temperature of the diode.

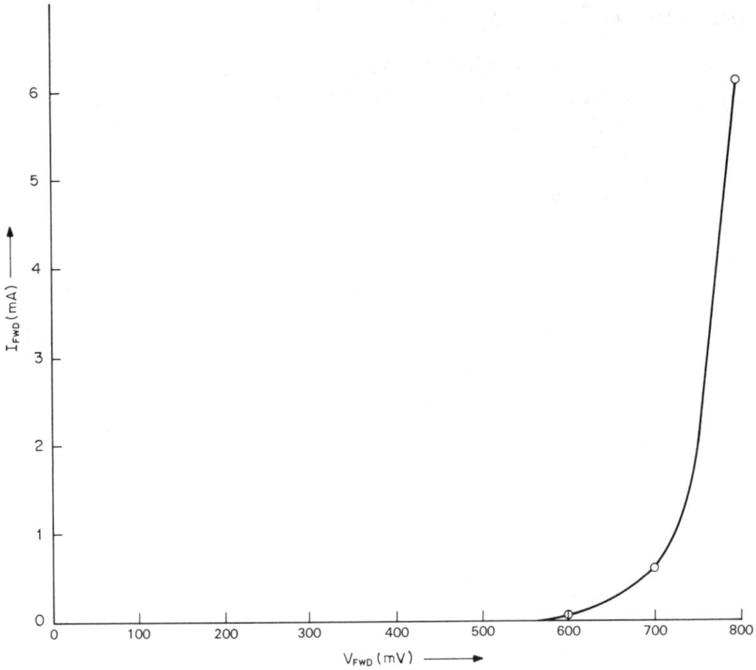

Figure 4.72 The forward characteristics of a silicon diode.

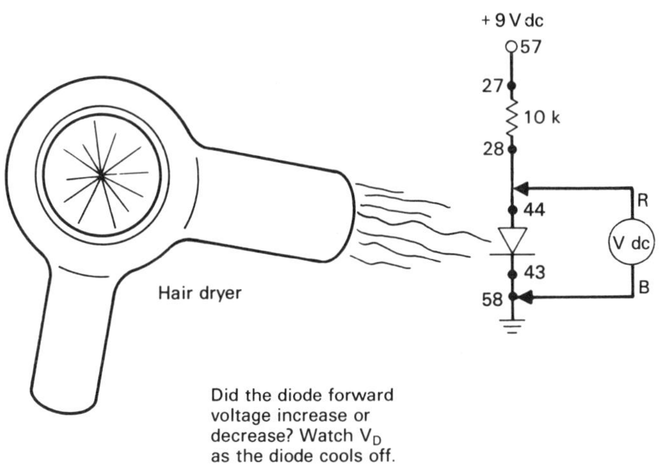

Figure 4.73 Measuring the effects of temperature on the forward voltage of a silicon diode.

The reverse characteristics of diodes

When a diode is reverse biased, large values of current no longer flow. An experiment to measure this reverse current is shown in Figure 4.74. If a silicon diode is used, the reverse current would be too small to measure with this meter.

If the germanium diode in the lab kit is used, the reverse current will be found to be approximately 25 μA. Now if the hair dryer is used (keep it 2 or 3 in away) and the germanium diode is heated while it is under reverse bias, the reverse current will increase. The forward voltage decreases and the reverse leakage current increases when the temperature of a diode is increased.

Transistors: The Three-Terminal On-Off Amplifying Devices

Now that diodes have been considered, some experiments have been performed, and the operation of diodes has been determined, we will turn our attention to something a little more complicated, a "bipolar transistor." There are basically two types of transistors available in electronics. One general type is the bipolar transistor, which will now

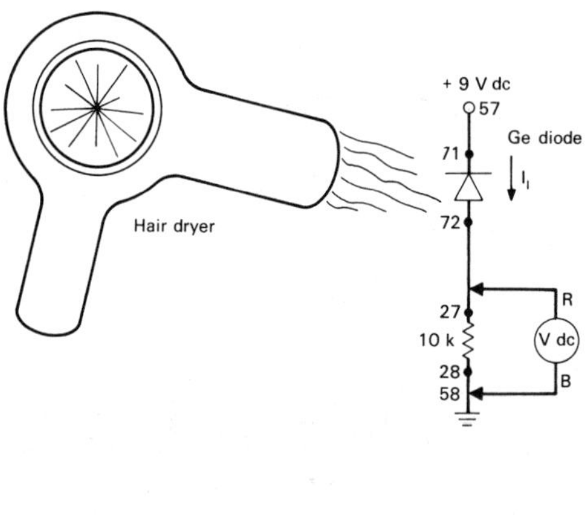

$$I_l = \frac{V}{10\ k}$$

V (at room temp) = 0.257 V

$$I_l = \frac{0.257\ V}{10\ k} = \frac{2.57 \times 10^1}{10^4}\ A = 2.57 \times 10^{-5}\ A$$

$$I_l = 25.7\ \mu A$$

V (when hot) = 0.532 V

$$I_{l\,(hot)} = \frac{0.532\ V}{10\ k} = \frac{5.32 \times 10^{-1}}{10^4}\ A = 5.32 \times 10^{-5}\ A$$

$$I_{l\,(hot)} = 53.2\ \mu A$$

Figure 4.74 Measuring the leakage current of a germanium diode.

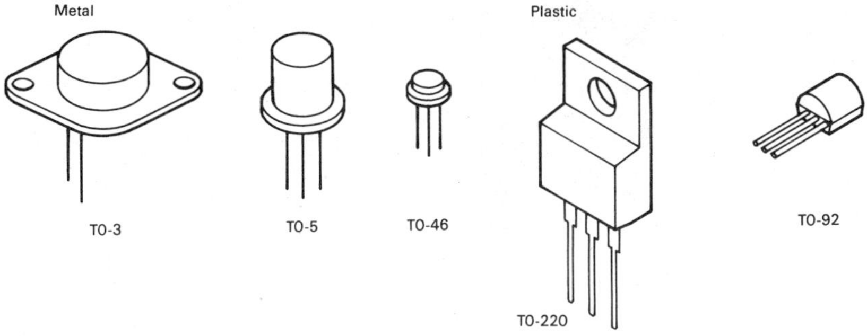

Figure 4.75 Some common transistors look like this.

be considered. A different type is the "field-effect transistor." Field-effect transistors aren't available in the lab kit and therefore won't be used in experiments.

Bipolar transistors are supplied in many different packages. Figure 4.75 shows the large metal TO-3 package (TO is an abbreviation for Technical Outline, the mechanical drawing for the package), smaller metal "cans" like the TO-5 and the TO-46, and some of the plastic packages that are used for transistors. The metal packages are generally supplied for high-operating-temperature applications. Plastic packages are used for low-cost systems.

Similar to a resistor, transistors can get hot, and some applications require a large package because the package is used to keep the transistor cool. In many applications there are such small voltages and currents involved that the power that is being dissipated in the transistor is not a concern. Near the output stages of an electronic system, like the output stage of a hi-fi power amplifier (that is providing the audio output power for a loudspeaker), large transistors are used that have to be bolted to special metal "heat sinks." In low-power circuitry, heat sinking of the transistors is usually not needed. Therefore, both small-sized (the physical size) transistors, and large-sized transistors are generally used in an electronic system. A large transistor (like a TO-3) indicates that there is a lot of power associated with that section of the system.

In the center region of the lab kit there are four transistors (as shown in Figure 4.76) that are available for experiments. The two on the left side have the symbol for a PNP transistor. Notice that the arrowhead (which indicates the "emitter") is pointing into the bar (the symbol for the "base"). The two transistors on the right have the arrowhead reversed: the arrow is pointing away from the bar for these NPN transistors.

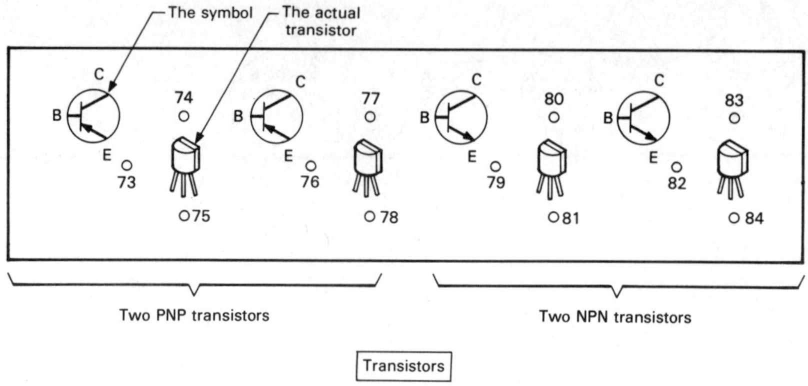

Figure 4.76 The four transistors in the lab kit look like this.

The small plastic TO-92 package is used. These transistors are therefore not capable of much power dissipation. In the experiments this will not be a problem because only relatively small voltages and currents will be used.

In electronic schematic diagrams, the symbol for a transistor should be drawn so that the emitter arrow points down the page. The emitter current then always flows from the top to the bottom of the page. This convention makes schematics easier to follow. The symbol for an NPN transistor should therefore be drawn with the emitter at the bottom, and a PNP transistor should be drawn with the emitter at the top, as shown in Figure 4.77.

The three leads of a transistor are the base, emitter, and collector. On the NPN transistor of Figure 4.77*a,* the base current and the collector current are seen to enter the transistor; the emitter current leaves. This means that the sum of the base current and collector current must equal

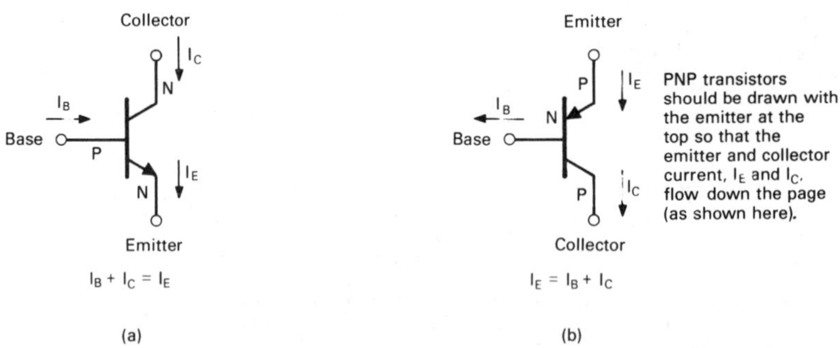

Figure 4.77 Symbols for bipolar transistors. (*a*) An NPN transistor and (*b*) a PNP transistor.

the emitter current; that is, the current *going in* must equal the current *coming out* (this basic relationship will be used later on).

A PNP transistor is shown in Figure 4.77*b*. Notice the direction of the currents for this transistor. Also notice that the emitter has been located at the top of the symbol. The emitter current, therefore, flows down, from the top of the paper to the bottom.

The two diodes in a transistor

There are two diodes associated with a bipolar transistor, as shown on the PNP transistor of Figure 4.78. One of these diodes is the emitter-base diode. Notice that the emitter-base symbol is very close to a diode symbol. The arrowhead indicates the forward direction for current conduction in this diode.

The other diode is the "collector-base diode." It is indicated with a dotted arrowhead in Figure 4.78*a*. This arrowhead is usually not shown so that there will be no confusion as to which lead is the emitter. Both the emitter and the collector of any bipolar transistor are always made of the same dopant type. In this PNP example, that dopant is P type.

The forward voltages of these diodes can be measured as indicated in Figure 4.78*b*. The 100-k resistor is used to limit the diode current flow. To the left is the circuitry to measure the voltage of the emitter-base diode, and to the right is the circuitry to measure the collector-base diode. These diodes are very similar to the silicon diodes that are available in the lab kit.

These two diodes can be drawn in a different way, as shown in the PNP transistor of Figure 4.79. This drawing uses the same cathode sym-

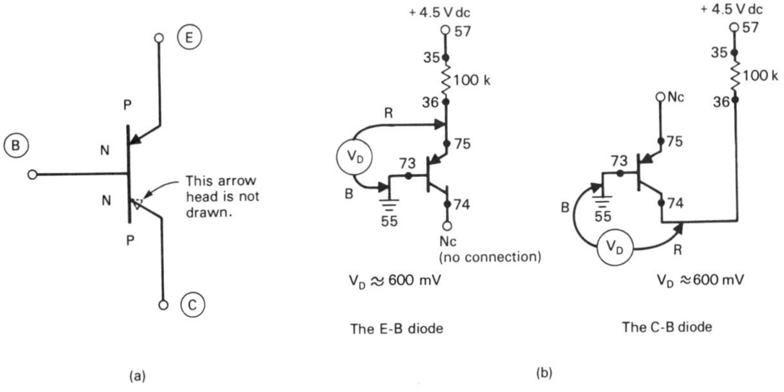

Figure 4.78 The two diodes in a transistor. (*a*) The PNP transistor and (*b*) measuring the forward voltage drop of each diode.

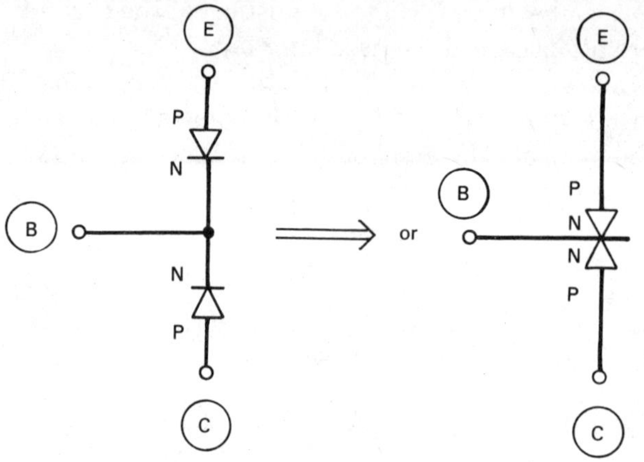

Figure 4.79 Drawing a PNP transistor as two back-to-back diodes.

bol and locates the anodes on both sides. Some originally unexpected action between these two diodes ("transistor action") is the key to bipolar transistor operation.

Demonstrating transistor action

Transistor action can be demonstrated by hooking up the test circuit shown in Figure 4.80. In Figure 4.80*a*, notice that the base lead is grounded. The emitter lead of the transistor is supplied with a forward current. This current is established by the 2.2-kΩ resistor that is connected to the 4.5-V power supply. The emitter current is about 1.8 mA.

With the multimeter set up as an ampmeter (tied between the negative 4.5-V power-supply voltage and the collector), the collector current can be measured. Notice that *the collector current is essentially equal to the emitter current.* Almost all of the emitter current appears in the collector. This is "transistor action." Transistor action takes place when the collector "steals" essentially all of the emitter current away from the base. As a result, the base current is very small (it is the difference between these currents, $I_B = I_E - I_C$).

In Figure 4.80*b*, the two diodes that are within a transistor are replaced with the two discrete silicon diodes that are in the lab kit. A similar demonstration circuit is used to see if transistor action takes place when using these two *separated* diodes. The same current is entered into one of the diodes and the common cathode point is similarly grounded. The ampmeter is connected in series with the lower

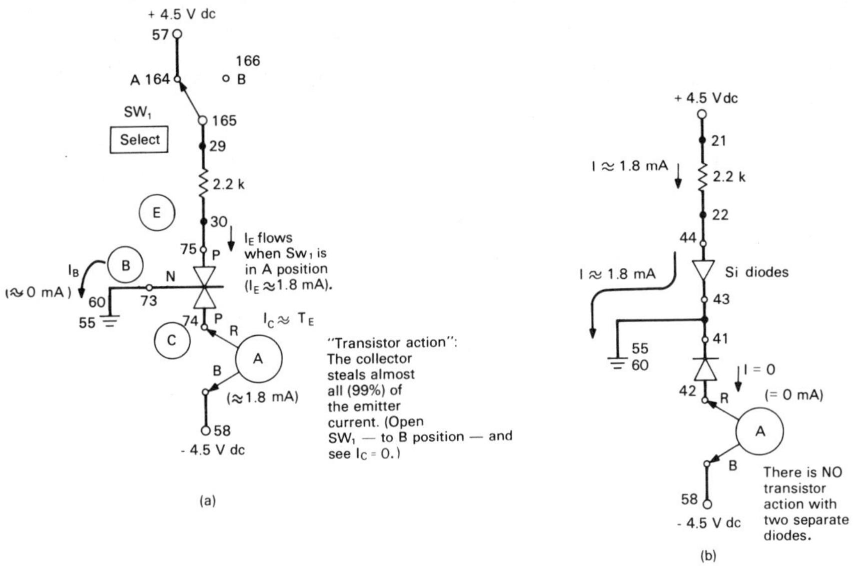

Figure 4.80 Demonstrating transistor action. (*a*) Transistor action and (*b*) using two separate Si diodes there is no transistor action.

diode to see if current flows (this would indicate the existence of transistor action).

Notice, in this case, that the ampmeter reads zero, indicating that none of the forward current of the upper diode is "stolen" by the lower diode. Therefore, transistor action does not exist between two separated diodes. The junctions of these diodes are too far apart. In fact, these junctions exist in separate packages. So, *transistor action occurs only when two diodes are built very closely together.*

Current gain (β) of a transistor

A small base current can control a much larger collector current. This can be demonstrated with the experimental setup indicated in Figure 4.81. The 1.5 V that is available from a single cell in the battery section is used, and a 100-kΩ resistor limits the current flow into the base lead. The base-emitter junction is a silicon diode, so the base lead will be about 600 mV above ground. The base current that flows can be determined by using the multimeter as a dc voltmeter to measure the voltage drop across the 100-kΩ resistor. This indicates about 10 μA of base current. If the multimeter is then connected as a dc ampmeter and put in the collector lead (as shown in Figure 4.81*b*), about 2 mA of current can be measured. This demonstrates the current gain of the

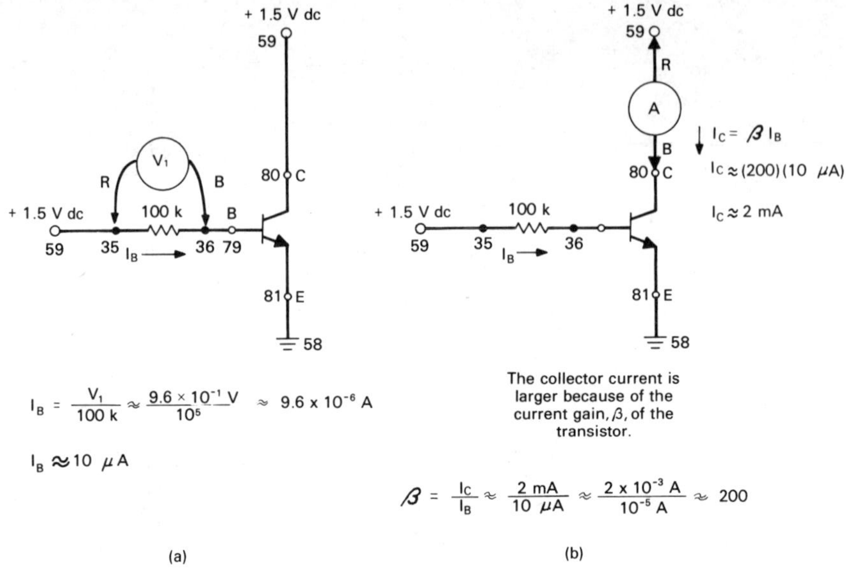

Figure 4.81 A small base current causes a large collector current. (*a*) Measuring the base currents and (*b*) measuring the larger collector current.

transistor. The 10 μA of base current is multiplied by the current gain of 200 to become 2 mA of collector current. A β of 200 is a very good value for a transistor.

A diode-connected transistor

A three-terminal transistor can be connected like a two-terminal diode. This is used many times in integrated circuits and it is called a "diode-connected transistor." Transistor action is still taking place and can be demonstrated in the diode-connected transistor shown in Figure 4.82.

A 4.5-V power supply and a 2.2-k resistor are used. The base lead is connected to the collector lead through a 10-k base-current sampling resistor. The meter is used to indirectly measure the current flow in the collector lead, as shown in Figure 4.82*a*. Essentially the full 1.7 mA of current flows in the collector lead. Now if the meter is used to indirectly measure the base current, Figure 4.82*b*, the base current can be seen to be very small, something like 9 μA. The base current is smaller than the collector current by a factor of 200:1. Even though the base is shorted to the collector, transistor action still takes place.

This diode-connected transistor has a forward voltage magnitude that is very similar to the forward voltage of a silicon diode. If this voltage is measured at room temperature and the transistor is heated using a hair dryer, as shown in Figure 4.83, this forward voltage is seen to drop (also at the rate of about 2 mV/°C).

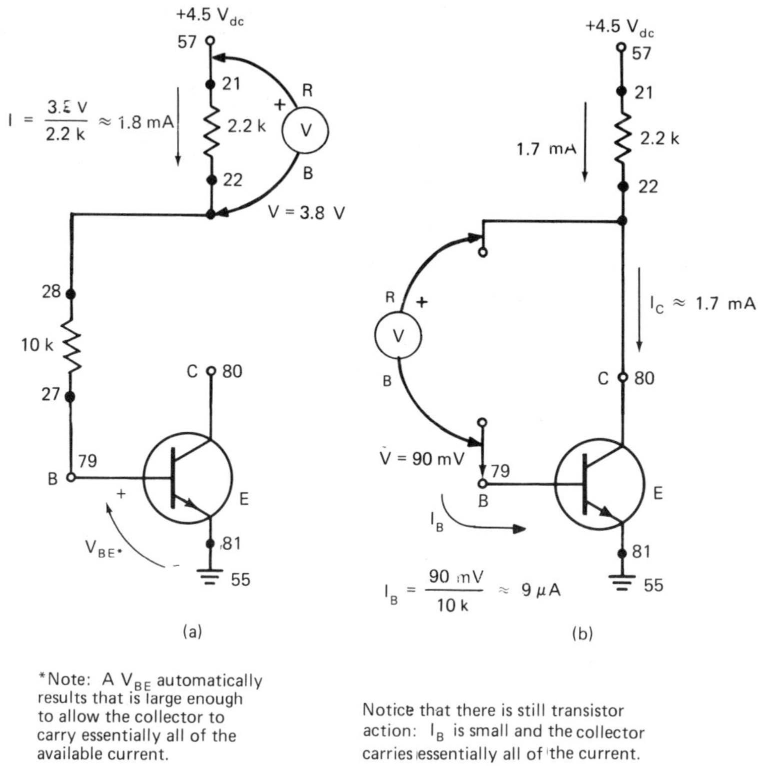

Figure 4.82 A diode-connected transistor. (*a*) Measuring I_C and (*b*) measuring I_B.

The Darlington connection

A useful connection, discovered by Sidney Darlington of the Bell Telephone Laboratories, is to tie two transistors together so that the overall current gain is much larger. This "Darlington connection" is shown in Figure 4.84. If 10 μA of current is fed to the base lead of this composite transistor structure, a multiplication by the current gain of the first transistor, Q_1, provides the emitter current. This larger current becomes the base current for the second transistor, Q_2. Therefore, the 10-μA input current is first multiplied by 200 to provide 2 mA. Then this 2 mA is again multiplied by 200 to provide 400 mA in the collector of Q_2. This Darlington transistor connection therefore functions like a single transistor that has a current gain that is the product of the individual current gains of the two transistors. In this example this product is 200 times 200, or 40,000. This means that a very small input current can be used to control a very large output current.

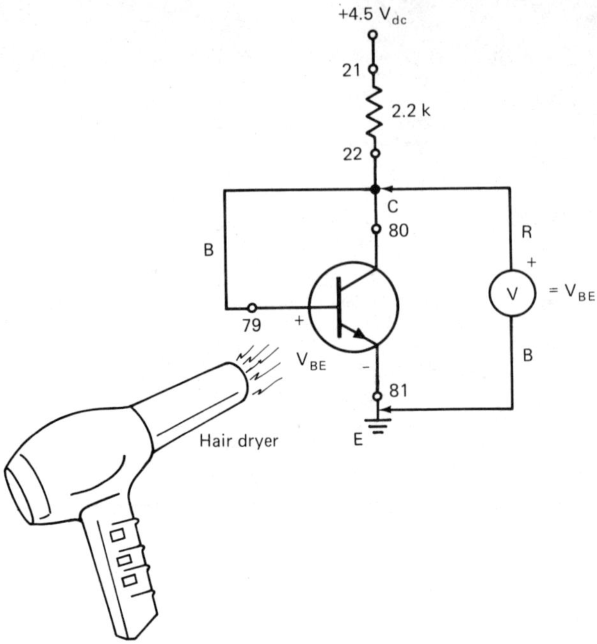

Figure 4.83 Measuring the effects of temperature on V_{BE}.

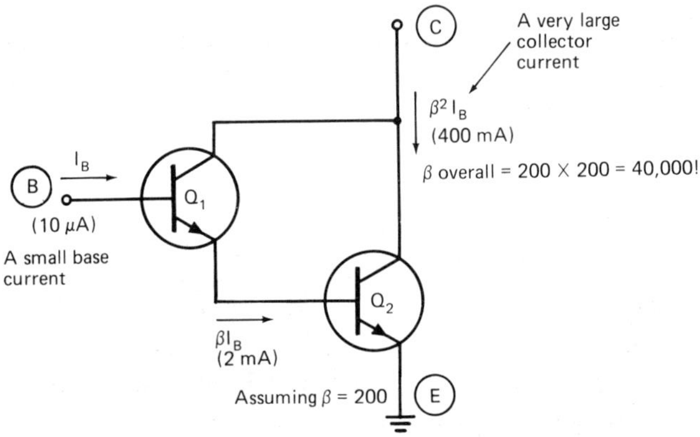

Figure 4.84 The Darlington connection.

Transistor collector characteristics

Special equipment is available to determine the electrical characteristics of transistors by displaying the relationship between the collector current and the collector-emitter voltage (for various values of base current) on the face of a cathode-ray tube (similar to a television but

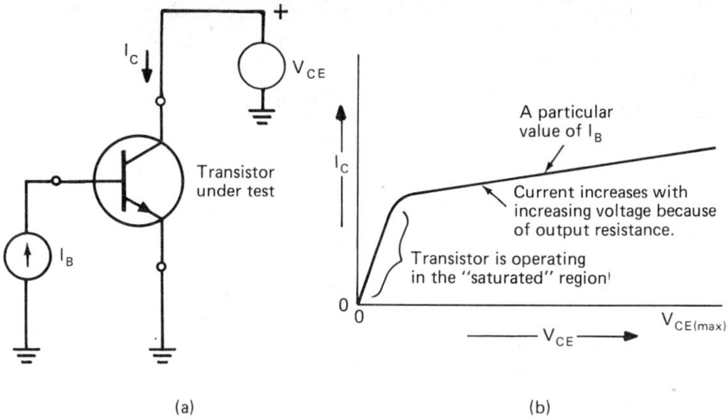

Figure 4.85 The idea of transistor collector characteristics. (*a*) The basic test setup and (*b*) the plot of the data obtained.

much smaller in size). The basic circuitry of a transistor curve tracer is shown in Figure 4.85. A sequence of increasing values of base current steps are established in the transistor; and, for each value of base current, the collector voltage is swept from zero to some adjustable high voltage. The collector current that results from each base current step is displayed as shown in Figure 4.85*b*. A collection of these traces is called the collector characteristics of the transistor. Each of the individual curves initially rises, as shown in Figure 4.85*b,* then it bends over (a "knee" of the curve), and then it rises again but much more gradually. In the region of the initial rise (where the collector-emitter voltage is small), the bipolar transistor is said to be operating in its "saturated region."

As the curve continues beyond the knee, notice that as the collector-emitter voltage increases, the value of collector current also increases but at a slower rate. This is a result of the output resistance of the transistor, an undesired effect. It is most desirable to have the collector current of a transistor remain constant in this operating region (the "linear" region) as the collector-emitter voltage increases.

The curves that result from a number of different base current values are all simultaneously displayed. Therefore, a family of curves results, ranging from a low value of base current to sequential steps of higher values of base current. Many of the operating parameters of a transistor can be determined by taking measurements from this family of curves.

Constant current sources

A constant current source (often simply called a current source) is a two-terminal electronic device that is somewhat similar to a constant voltage source (often simply called a voltage source). Whereas an ideal

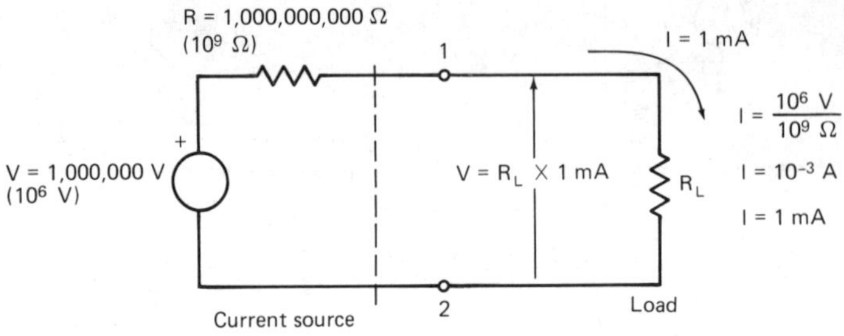

Figure 4.86 A 1-mA constant current source.

voltage source (such as a car battery or the 110-V electrical outlets in a house) would supply any amount of current that a circuit would require from it and still maintain the terminal voltage constant, a constant current source would supply a fixed, constant current that is independent of the voltage that results across the terminals of the current source.

The circuit shown in Figure 4.86 is a conceptual way to build a current source. It makes use of a 1,000,000-V voltage source and a 1,000,000,000-Ω resistor. If terminals 1 and 2 of this current source were shorted together, a current of 1 mA would flow. Notice that if a load resistor (R_L) were placed across terminals 1 and 2 of this current source, a current of 1 mA would still flow because the resulting voltage across the terminals would most likely be small compared to the internal voltage source of 10^6 V (and could therefore be neglected in its effect on current flow) and also this load resistor would most likely be small when compared to the internal resistance of 10^9 Ω (and could therefore also be neglected in its effect on current flow). This is an impractical way to build a current source and other circuits are therefore used.

A current source can be made using a PNP transistor, as shown in Figure 4.87. A biasing network, which consists of two silicon diodes, is used. The 10-k resistor turns on these two series-connected diodes. The base-emitter voltage of the PNP transistor is nearly the same as the forward voltage of one of these diodes. Therefore, approximately a diode voltage exists across the 47-k resistor in the emitter lead. This provides an emitter current of 892 mV/47 k, or 19 μA. Because the emitter and collector currents are essentially the same, the collector current coming out of terminal 74 of the PNP transistor will also be 19 μA.

Some measurements can be made to determine how constant this current remains as the collector voltage changes. This is done by using load resistors of various sizes, shown as R_L, that are tied between the collector and ground. By sequentially putting in place various load resis-

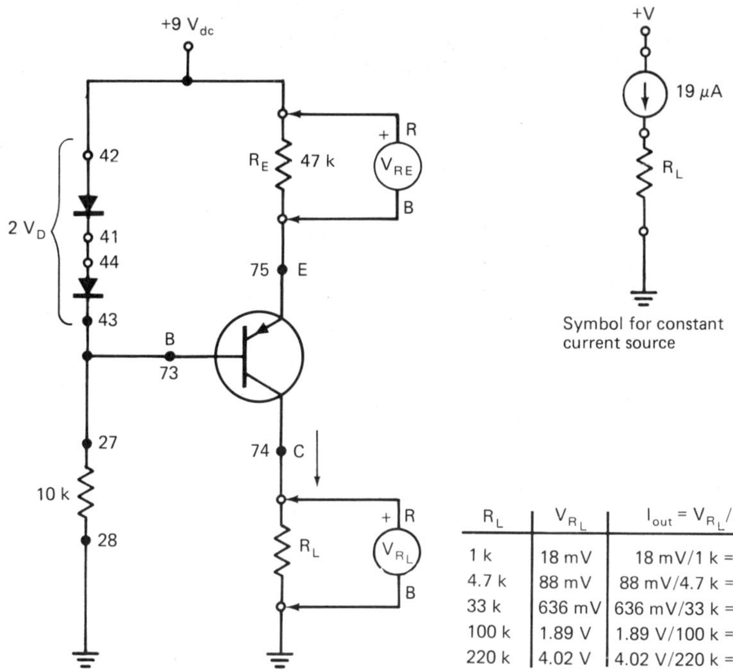

R_L	V_{R_L}	$I_{out} = V_{R_L}/R_L$
1 k	18 mV	18 mV/1 k = 18 μA
4.7 k	88 mV	88 mV/4.7 k = 18.7 μA
33 k	636 mV	636 mV/33 k = 21 μA
100 k	1.89 V	1.89 V/100 k = 18.9 μA
220 k	4.02 V	4.02 V/220 k = 21.4 μA

Figure 4.87 Biasing a PNP transistor to provide a constant current source.

tors and then measuring the voltage drop across each of these resistors, the output current can be calculated, as shown in the chart of tabulated data that is underneath the schematic diagram. Notice that over a wide range of load resistance values essentially the same amount of current flows out of the collector of this PNP transistor current source.

One use for a constant current source is to provide a linear voltage sweep. This results when this constant current is caused to flow into a capacitor, as shown in Figure 4.88. Again, we use the Key switch (and a 100-Ω resistor) connected across the capacitor to initially remove the charge from the capacitor and to therefore start with a sweep voltage of zero across the capacitor. When the Key switch is opened, the sweep voltage will then slowly, and very linearly, increase. This sweep voltage will increase until it gets to within about 1 V of the positive power supply voltage. The voltage won't go clear to 9 V because the collector of the PNP transistor can't go to a voltage that is larger than its emitter voltage. The emitter voltage is 892 mV less than the 9-V power supply voltage, so this increasing ramp should stop at a voltage slightly larger than 8 V.

With a current flow of 19 μA into the 100-μF capacitor, this total linear voltage sweep will take about 42 s. At the end of the sweep, the

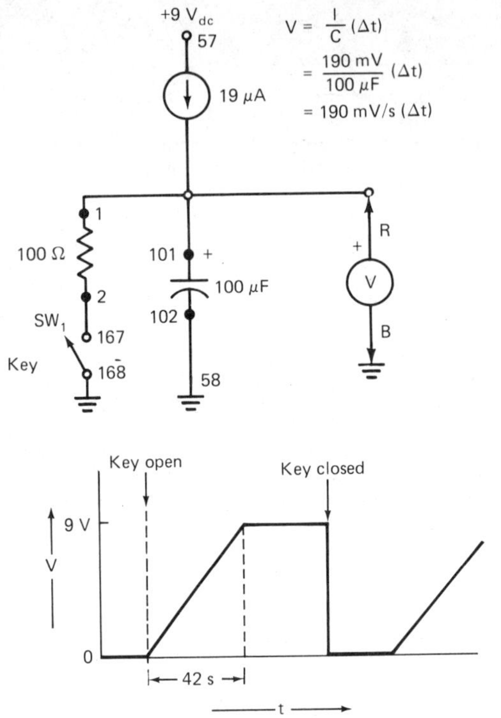

Figure 4.88 Making an increasing-voltage sweep circuit.

transistor will saturate and the sweep voltage will remain at 8 V. When the Key switch is again closed, the voltage across the capacitor is reset to zero so that the sweep voltage returns to 0 V.

An NPN transistor can also be used to make a current source, as is shown in Figure 4.89. A constant current is now available out of the collector of this transistor. Again, a diode voltage drop exists across the 47-k emitter resistor. Therefore, 19 μA of current will flow in the collector of this NPN transistor. Load resistors now are connected to the positive power supply voltage. If the voltage drop across these resistors is measured, we can determine how constant the current magnitude actually is.

A linear voltage sweep circuit, which starts at the positive power supply voltage (9 V) and then decreases with time, is shown in Figure 4.90. Now, when the Key switch is closed, the voltage across the capacitor will still discharge to zero, but the sweep voltage will jump to 9 V. When the Key switch is opened, a decreasing linear voltage sweep results that reaches 1 V in about 42 s.

There are many uses for linear sweep voltages in electronics. For

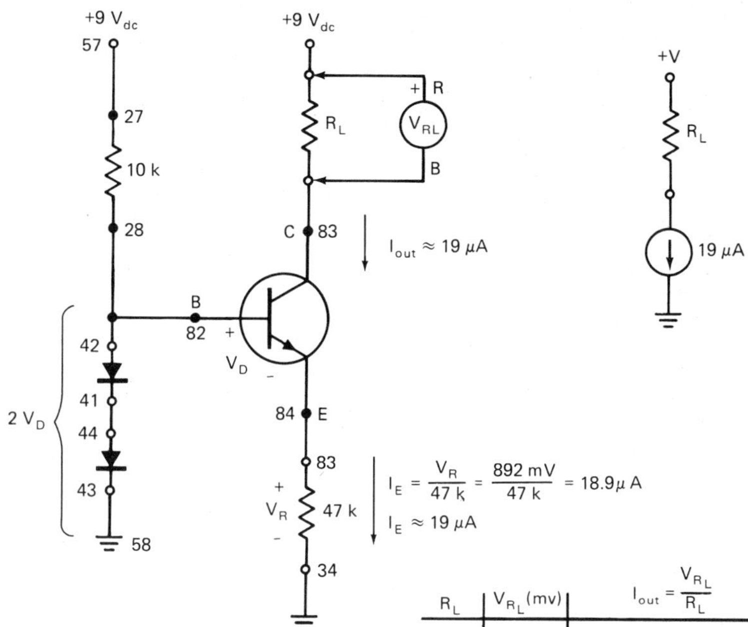

R_L	V_{R_L} (mv)	$I_{out} = \frac{V_{R_L}}{R_L}$
1 k	18	18 m V/1 k = 18 μA
4.7 k	87	87 m V/4.7 k = 18.5 μA
33 k	631	631 m V/33 k = 19 μA
100 k	1.875 V	1.875 V/100 k = 18.2 μA
220 k	4.00 V	4.00 V/220 k = 18.8 μA

Figure 4.89 Biasing an NPN transistor to provide a constant current source.

example, a ½-min time delay can be obtained by the time that elapses until the voltage across the capacitor increases from 0 to 5.7 V. A voltage comparator could detect this end-point voltage. (Voltage comparators operate with a reference voltage input and, when a second input voltage equals this reference voltage, the output of the voltage comparator changes from ON to OFF or vice versa.) Voltage sweep circuits are used to provide a voltage analog of time. As time increases, a linear sweep voltage can be made to either increase or decrease in voltage.

The base-emitter junction in breakdown

In a typical silicon diode, the reverse breakdown voltage rating is purposely kept very high so that under no conditions within the operating range of the circuit will the diode enter breakdown. Other diodes are available that are designed to be operated continuously while in reverse breakdown. These breakdown diodes are called "zener" diodes and "avalanche" diodes. Actually, zener diodes exist with breakdown volt-

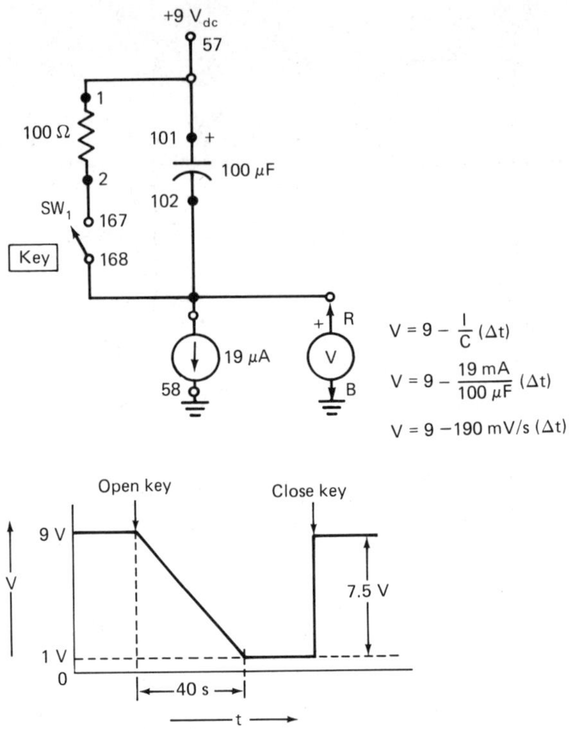

Figure 4.90 Making a decreasing-voltage sweep circuit.

ages of about 2 V up to about 4 V. In the range of 6 V up to as high as 200 V, the avalanche breakdown mechanism takes place. In the cross-over range, about 4 to 6 V, both breakdown mechanisms take place simultaneously within the diode.

The base-emitter junction of a modern bipolar transistor will not sustain large values of reverse voltage. In fact, we can break down the base-emitter junctions of the transistors that are supplied in the lab kit to demonstrate avalanche diodes. They are called "avalanche" because the breakdown voltage is relatively large, greater than 4 V.

We will use the PNP current source that has been previously described to limit the current flow during the breakdown of the emitter-base junction of one of the NPN transistors, as shown in Figure 4.91. To prevent permanent damage to the transistor, it is important to use a small magnitude of current and to spend only a short time in this reverse breakdown condition. Therefore, the Key switch is used to break the diode down only long enough to obtain a voltage reading. Operating the base-emitter junction of a bipolar transistor in breakdown can cause a degradation in the current gain, especially with high current flow and at high temperatures.

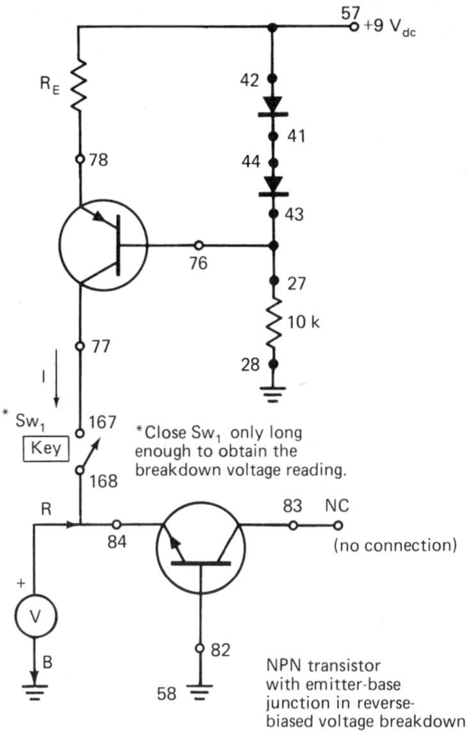

(a)

R_E	I	V
470 k	1.3 μA	8.40 V
100 k	6 μA	8.41 V
10 k	60 μA	8.40 V
1 k	600 μA	8.42 V

Voltage regulator action

With a wide range of current — get a constant value (or a reference value) of voltage.

(b)

Figure 4.91 The base-emitter junction as an avalanche breakdown diode. (*a*) The test circuit and (*b*) the measured data.

Notice that the size of the emitter resistor (R_E), is changed to vary the magnitude of the current that flows into the base-emitter junction. In the tabulated data, shown in Figure 4.91*b*, you can see that although the breakdown current through the diode was increased from a little over 1 μA to 900 μA, there is not much of a change in the magnitude of the breakdown voltage of this avalanche diode. This provides a very good voltage regulating action, and both avalanche and zener diodes are used as voltage references because the breakdown voltage is fairly predictable and constant, even though the magnitude of the biasing current might change.

A breakdown diode is used in the simple power supply voltage regulator shown in Figure 4.92. In electronic systems, problems exist because the ac power line voltage has a nominal value of 110 v, which means this voltage can have values less than and also greater than 110 v. In electronic system design, it is desirable to establish a constant-voltage dc power supply for the electronic circuits, even though the

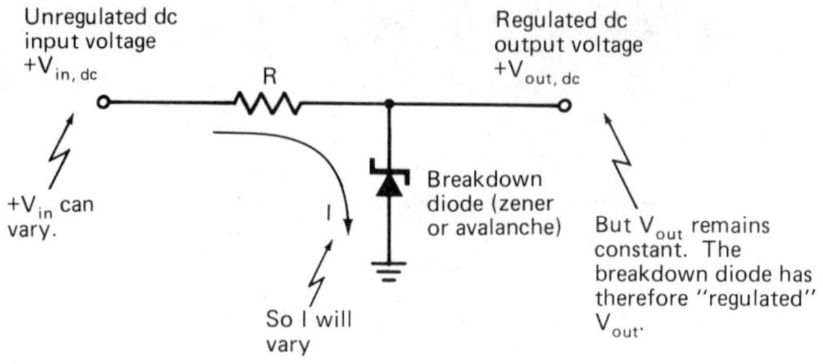

Figure 4.92 A shunt voltage regulator.

power line voltage and the unregulated dc voltage out of the rectifier circuitry may wiggle around.

In the shunt voltage regulator of Figure 4.92, this unregulated dc input voltage is applied to the resistor R. As the unregulated input voltage changes, the current flow through the diode will change, but the output voltage will remain essentially constant. Therefore the output voltage is regulated. This shunt regulator is one of the simplest voltage regulators.

The effects of temperature on the base-emitter breakdown voltage

The effects of increasing temperature on the magnitude of the breakdown voltage of the base-emitter junction of the NPN transistor can be determined by measuring the magnitude of the breakdown voltage at room temperature and then heating the transistor with a hair dryer, as shown in Figure 4.93. Notice that as the transistor is heated, the breakdown voltage increases slightly, indicating a positive temperature coefficient.

A transistor as a dc voltage amplifier

A transistor can be used to amplify dc voltages. This is shown in the test setup of Figure 4.94. This circuit allows the dc input voltage directly on the base of the transistor to be measured as well as the dc output voltage at the collector. By varying the Control, the dc input voltage can be varied to control the dc output voltage. To obtain these data, adjust the Control so as to provide the prescribed dc output voltages that are listed in the left-hand column of the table on the figure. Start with a dc output voltage of 0.1 V and then increase this up to a maximum of 4.5 V. Measure what dc input voltage, directly at the base, was necessary in order to obtain each of these output voltages.

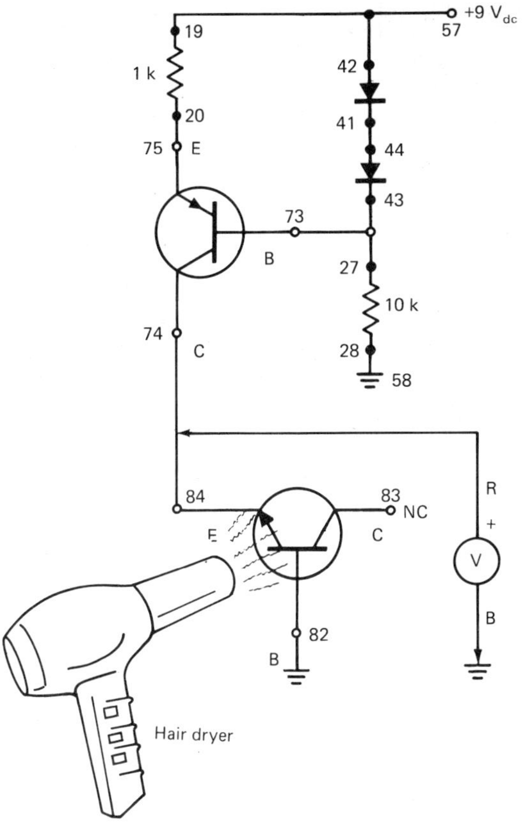

Figure 4.93 The effects of temperature on the base-emitter breakdown voltage.

For example, to obtain a dc output voltage of 0.1 V, 567 mV is needed directly at the base of the transistor. Notice that as the dc output voltage increases, the dc input voltage decreases. This is called "inverting" gain and results with a common-emitter amplifier (the emitter is tied to ground and is the common terminal for both the input and output voltages). This inversion takes place because as the base voltage increases, more collector current flows through the 100-k collector load resistor R_c, which causes the collector voltage to decrease.

Notice from these data that to cause the output voltage to make a large change, from 1 to 4.5 V, requires that the input voltage decrease from 553 to 450 mV. Or, the input voltage only has to change 103 mV in order to cause a 3.5-V change in the output voltage. The ratio of this change in output voltage to the change in input voltage is called the "large signal" voltage gain. In this example, the large signal voltage gain of this single-transistor amplifier is 3.5 V divided by 103 mV, or 34.

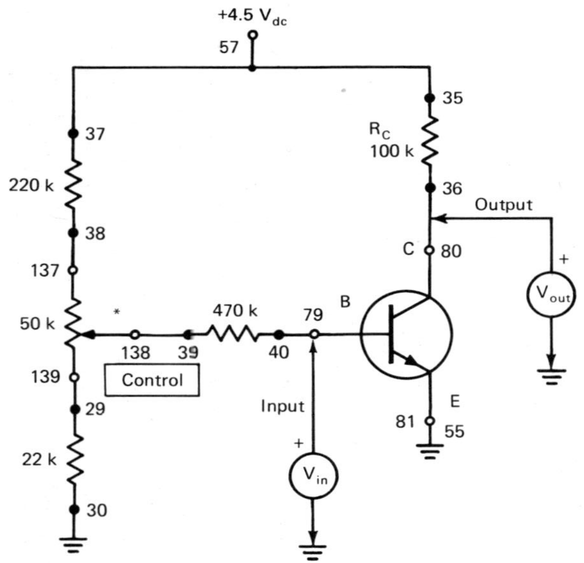

*Adjust control to obtain each value of V_{out} listed; then read and record V_{in}.

Output V_{out} (volts)	Input V_{in} (mV) (to obtain the listed value of V_{out})
0.1	567
0.3	558
0.5	557
1	553
2	542
3	527
4	497
4.5	450

Note that the output voltage goes up as the input voltage goes down—an inverting amplifier.

553 mV
−450 mV
103 mV

As an example of gain:

An increase in V_{out} of $(4.5 - 1) = 3.5$ V → Required a decrease in V_{in} of only 103 mV

So this large-signal voltage gain A_V is

$$A_V = \frac{\Delta V_{out}}{\Delta V_{in}} \approx \frac{3.5 \text{ V}}{103 \text{ mV}} \approx \frac{3.5 \text{ V}}{1.03 \times 10^{-1} \text{ V}}$$

$$A_V \approx 34$$

Figure 4.94 The test setup to demonstrate a single-transistor dc voltage amplifier.

A transistor as an electronic switch

In digital circuit applications, the transistors are simply turned on or off. The transistors are used as electronically controlled switches. An example of a transistor used as a switch is shown in the simplified representation of Figure 4.95. In Figure 4.95*a*, the collector-emitter circuit of the transistor has been replaced with a switch. This switch is controlled by the magnitude of the base voltage. With a base voltage of zero, there is no base current. The transistor is off and the switch is open. Because the switch is open, the output voltage will be the full power-supply voltage (5 V dc). That is one of the two possible states of the output voltage. This represents a logical 1, the high-voltage state.

If a large control voltage, say 5 V, were applied (an input resistor is used to limit the base current), the transistor would be turned on. This conducting transistor would be in saturation (operating just as if a switch between the collector and emitter leads were closed). A short circuit now exists between the collector and the emitter. This ties the collector to ground, so the output voltage is 0 V. An output voltage of zero is called a logical 0.

A primitive *RS* flip-flop

An example of a logic circuit that has electronic "memory" is the direct-coupled Reset-Set (*RS*) flip-flop shown in Figure 4.96. This circuit is easy to construct and is relatively easy to understand. This flip-flop uses the two NPN transistors that are available in the lab kit. Notice

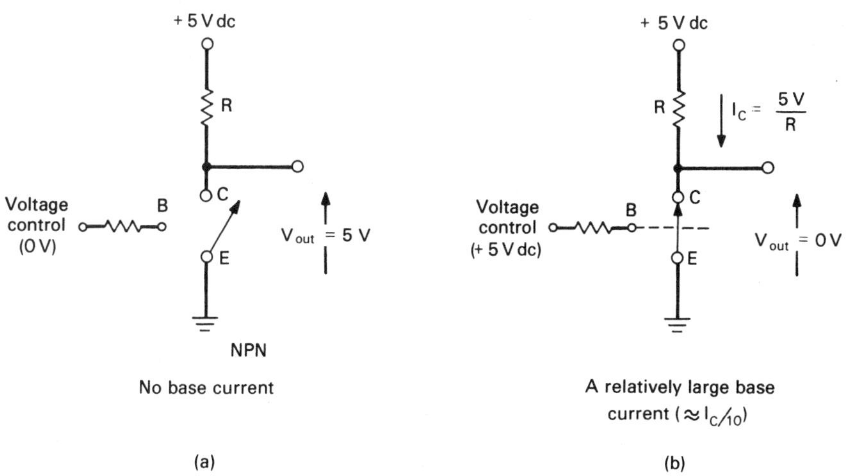

Figure 4.95 A transistor as a switch. (*a*) Transistor (switch) off and (*b*) transistor (switch) on.

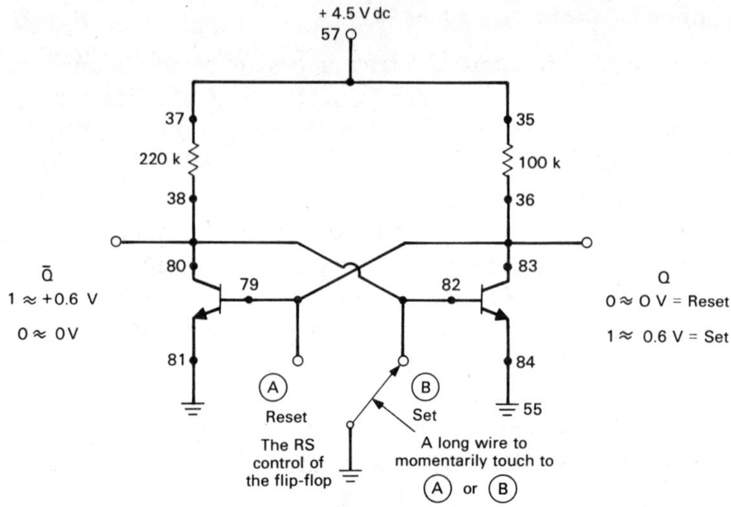

Figure 4.96 A primitive *RS* flip-flop using direct-coupled transistor logic (DCTL).

the crossing of the wires in the center of the circuit diagram. The collector of one transistor ties over to the base of the other, and vice versa. This "cross coupling" is characteristic of a flip-flop circuit.

If the collector output on the left is arbitrarily called $\overline{Q}$ (pronounced "Q bar") and the collector output on the right is called Q, whatever signal exists at Q, the complement or the opposite logic state always exists at $\overline{Q}$. The bar over the symbol Q implies the complement of whatever logic state Q is in. For example, if the transistor on the right side is conducting, the voltage at Q will be approximately 0 V (representing the *reset state* of this flip-flop). Zero volts at the collector of the right-side transistor means that the base voltage for the transistor on the left will also be zero. With 0 V on the base, the left-side transistor will be off, so this collector voltage will try to rise to 4.5 V.

Notice that the collector of the left transistor ties back to the base of the right transistor. That base (also called point B) is the "Set input." Therefore, as the collector voltage attempts to move up to 4.5 V, the base-emitter diode of the other transistor catches it and limits it to about 600 mV. When the Q output is zero, the $\overline{Q}$ output is in the logical 1 state at 600 mV.

If the Set input is momentarily grounded (with a wire that is touched to ground), the base of the right-side transistor goes to zero (which shuts off that transistor). Therefore, the collector voltage of this transistor rises. The base of the left-side transistor (the point marked A) clamps this rising voltage at one diode drop (600 mV). This turns on the left-side transistor. The wire that was used to set the flip-flop is therefore no longer needed, and this logic state will remain as long as the power-supply voltage is applied. Therefore, a flip-flop has memory. It remembers whatever logic state it was forced into. This flip-flop can be set or reset by shorting alternate bases to ground. In the Set state, the Q output is a logical 1 (600 mV) and the $\overline{Q}$ output will be a logical 0 (0 V). This RS flip-flop is called a "bistable circuit" because it has two stable states.

The astable multivibrator

An "astable" multivibrator circuit has no stable states. It constantly changes from one state to the other: it "oscillates." An astable multivibrator is sometimes used as a clock circuit for a simple digital system.

An astable multivibrator can be built using the components that are in the lab kit, as shown in Figure 4.97. The two 100-μF capacitors provide a low-frequency oscillation that can easily be observed using the dc voltmeter. The time that this circuit remains in each state can be determined by counting seconds.

The cross coupling is now provided by capacitors. This is the reason that stable states are not achieved. These capacitors (C_1 and C_2) are

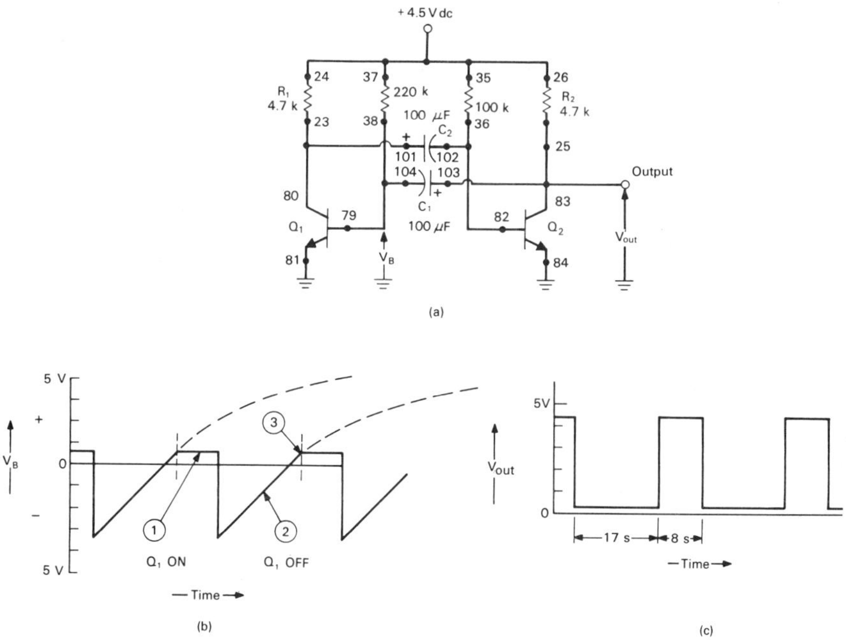

Figure 4.97 An astable multivibrator. (*a*) The circuit diagram, (*b*) the V_b waveform, and (*c*) the output waveform.

the key to the circuit action. Initially, assume that the transistor on the left (Q_1) is on. The current through the 220-kΩ resistor keeps it on. This is region 1 on the V_B waveform shown in Figure 4.97*b*.

While Q_1 is on, Q_2 will be off (because of a negative voltage that appears at the base of Q_2; this will be discussed shortly). With Q_2 off, C_1 is charged up to nearly the 4.5-V supply voltage by R_2. The left side of C_1 is tied to the base of Q_1, which provides a voltage of approximately 0.6 V, so C_1 will charge up to 4.5 V minus 0.6 V, or 3.9 V. The negative voltage at the base of Q_2 is slowly increasing because of the current flow through the 100-kΩ resistor. Eventually this voltage will become positive and large enough (0.6 V) to turn Q_2 on. This then causes a change of state.

When the circuit changes states, Q_2 goes on and Q_1 goes off. Notice that when Q_2 goes on, the right side of C_1 (being tied to the collector of Q_2) goes to essentially 0 V. The voltage stored on C_1 therefore causes the base of Q_1 (V_B on Figure 4.97*b*) to abruptly swing to a negative value (approximately −3.3 V, the region labeled "2" on the V_B waveform). It is this negative voltage at the base of Q_1 (or Q_2 for the opposite state) that keeps this transistor off.

During this quasi-stable state, the 220-kΩ resistor is providing a current flow into C_1 that is charging this capacitor toward 4.5 V. The negative plate of the electrolytic capacitor (C_1) is now on its way to +4.5 V. This voltage change on C_1 follows the exponential curve shown as region 2 of Figure 4.97*b*. When V_B increases to +0.6 V, Q_1 is no longer held off and therefore is turned on by the current coming from the 220-kΩ resistor. This is the point labeled "3" on the V_B waveform of Figure 4.97*b* and terminates that quasi-stable state.

In this astable multivibrator, one of the base voltages is always negative. This provides the quasi-stable states. Use the dc voltmeter to observe the V_B voltage and also the output voltage (V_{out}) shown in Figure 4.97*c*. This output voltage is not symmetric because unequal base resistors are used (100 and 220 kΩ). This provides the approximately two-to-one difference in the on-off timing. Equal-valued base resistors and equal-valued capacitors would provide a symmetrical square wave output voltage waveform.

The monostable multivibrator

A "monostable" multivibrator has *one* stable state. This circuit, shown in Figure 4.98, uses a resistor (22 kΩ) to cross couple on one side and a capacitor (100 μF) to cross couple on the other side. Therefore, this is a combination of a bistable with an astable circuit. The new thing is the trigger input that is used to initiate the quasi-stable state.

At rest, Q_1 is on and Q_2 is off. Momentarily closing the Key switch causes the voltage at the base of Q_1 to swing negative (as in the pre-

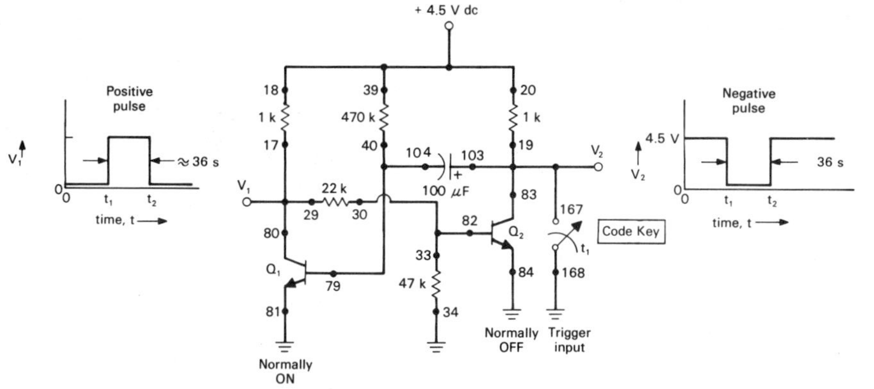

Figure 4.98 A "one-shot" pulse-forming circuit.

vious astable circuit). This negative voltage at the base of Q_1 turns Q_1 off, and the rising V_1 voltage couples through the 22-kΩ cross-coupling resistor to turn Q_2 on. Because Q_2 rapidly goes on, only a short-time-duration trigger is needed. This quasi-stable state persists while the 470-kΩ resistor charges the 100-μF capacitor and eventually causes the base voltage of Q_1 to increase to +0.6 V. At this time the circuit returns to its reset state. This produces an output pulse for each trigger input. This circuit is also known as a "one-shot multivibrator."

A toilet is a hydraulic analogy of this monostable multivibrator. The flushing lever is the trigger input. The flushing action continues until the local reservoir of water in the tank is drained, which ends the quasi-stable state. The system then "recharges" by again filling the tank and then returns to its stable rest state, awaiting the next input trigger.

The Silicon-Controlled Rectifier

A number of specialized semiconductor devices (in addition to diodes and transistors) have appeared over the years. One of the more successful of these is the silicon-controlled rectifier (SCR) that is used in power-control applications. A similar device, the "triac," is also available for ac power-control applications. The SCR will be described because undesired parasitic SCRs exist within digital ICs.

The symbol for an SCR is shown in Figure 4.99*a*. A sequence of three regions of alternating doping types is needed to make NPN or PNP bipolar transistors. If an additional layer is added, to provide an NPNP or PNPN structure, an SCR can be created. Therefore, an SCR can be considered as a special "hook connection" of a bipolar NPN and PNP transistor, as shown in Figure 4.99*b*.

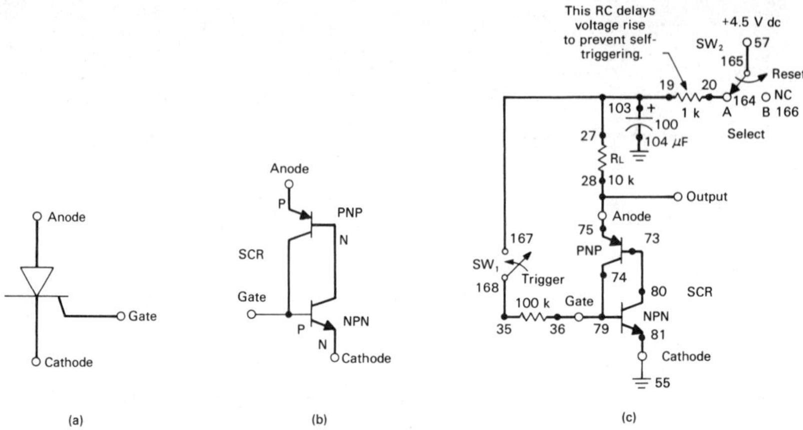

Figure 4.99 Making a silicon-controlled rectifier out of two transistors. (*a*) The symbol for an SCR, (*b*) the hook connection of two transistors to make an SCR, and (*c*) a circuit equivalent of an SCR.

This is a "regenerating" circuit. Initially, consider that current is somehow being supplied out of the collector of the PNP. This becomes base current for the NPN. The collector current of this NPN is larger because of the current gain of this transistor. This increased collector current flow is base current for the PNP. Again, this PNP base current is increased by the current gain of the PNP to provide a larger collector current. This reinforces the initially assumed current flow. If either transistor is momentarily turned on, this circuit will regeneratively "latch up" so that both transistors are conducting very large currents (limited *only* by the external circuitry).

An SCR can be built out of the NPN and PNP transistors that are available in the lab kit, as shown in Figure 4.99*c*. The 10-kΩ resistor (R_L) tied to the anode of this simulated SCR establishes the current limit when this SCR latches.

To trigger this SCR, the Key switch (SW_1) inputs an initial base current to the NPN transistor. The Key switch needs only a very quick tap because a short-duration current pulse is enough to get the rapid regeneration mechanism started. The 100-kΩ resistor in series with the Key switch limits this triggering base current flow.

An SCR, once latched, will stay in the latched-up mode. To reset this SCR, the Select switch (SW_2) has to momentarily remove the power by being switched to position *B*. The 1-kΩ and 100-μF RC network on the *A* side of this switch slowly reapplies voltage to the SCR. Remove the ground lead of this capacitor and notice that the abrupt application of anode voltage to the SCR will also trigger it on.

The output of the SCR (the anode) will be at +4.5 V until the SCR is triggered on. While on, the output voltage will be slightly larger than one forward diode drop (or about 650 mV).

Complex IC processes, such as all of the bipolar and the complementary MOS (CMOS) processes, provide adjacent regions in silicon that follow this NPNP or PNPN sequencing. Therefore, undesired parasitic SCRs become a problem in the applications of these ICs.

Special processing steps have been taken in some of the CMOS processes that eliminate this parasitic SCR problem. The simplicity of the wafer processing that was used for the older PMOS and the newer NMOS digital processes does not provide parasitic SCRs. Therefore, for MOS digital circuits, this SCR problem first appeared with the introduction of CMOS logic circuits.

We have just completed a look at electronic components when operated with dc voltages. Now we will shift our emphasis to see what happens when the voltages are no longer constant but instead change as a function of time. We will therefore next consider ac voltages.

Chapter

5

An Introduction to AC Circuits

The most common ac voltage is the electric power that is supplied to homes and industries. This voltage is not constant, like the voltage of a battery, but is cyclically rising and falling. It spends half its time positive and the other half negative and is always changing in a smooth way. Most linear circuits amplify and process voltages that are changing with time. These are incorrectly called ac voltages, which translates to "alternating current voltages," but this has been standardized and is accepted. You will never hear of "alternating voltages"—which is more correct. Because a large number of the ac voltages are sine waves, we will start by looking at these.

Sine Waves

Many examples of sine waves exist in nature. For example, a rock thrown into a pond of water will produce concentric ripples that move outwardly from the place the rock fell into the water. These outward moving ripples produce water ripples that are sine waves of decreasing amplitude. Large amplitudes exist in the vicinity of the point of impact and only smaller amplitudes remain at larger distances, as shown in Figure 5.1.

A child on a swing, Figure 5.2, or the pendulum of a clock also produce a sinusoidal motion as the object oscillates or moves back and forth.

If a child in a swing were to bend over with a pencil in his or her hand and traced his or her position versus time on a piece of paper that was slid underneath, the graph shown in Figure 5.3 would result.

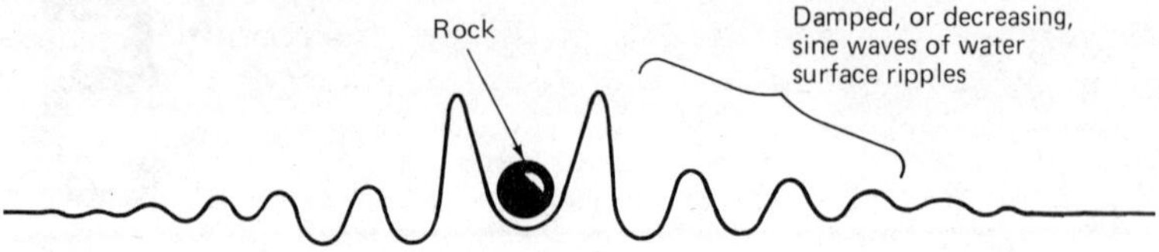

Figure 5.1 The sine waves produced by a pebble dropped into water.

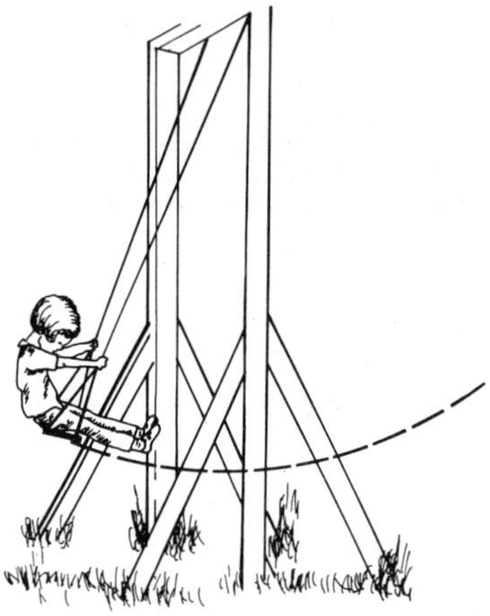

Figure 5.2 The sinusoidal motion of a child on a swing.

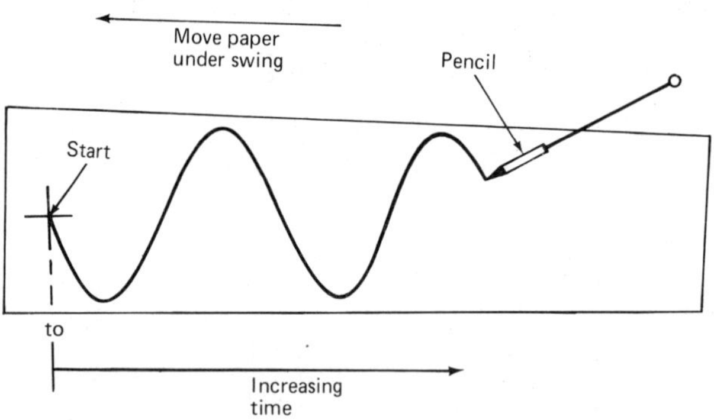

Figure 5.3 Marking the position of the swing on a strip of paper.

This is the very familiar sine wave—something that smoothly increases and then smoothly decreases and is constantly changing.

A few sine waves are shown in Figure 5.4. Figure 5.4*a* shows a relatively low-frequency sine wave, 1 Hz, or 1 cycle per second (the abbreviation Hz, in honor of Heinrick Hertz, is used to represent cycles per second). This voltage waveform starts at zero, increases to a maximum positive peak value, starts decreasing, again goes through zero, increases to a maximum negative value, and then goes back to zero. This completes one cycle; the waveform then repeats this same pattern. If one cycle is completed in 1 second (s) the frequency is 1 cycle per second, or 1 Hz.

Figure 5.4*b* shows a sine wave completing two cycles in this same time period of 1 s. This represents a frequency of 2 Hz. Figure 5.4*c* shows a waveform completing 10 cycles in the same 1-s time interval: 10 Hz. Frequency is one description of a sine wave, and it indicates how many cycles are completed in a 1-s interval of time.

In electronics, voltages and currents exist over a wide range of sinusoidal frequencies. The names and abbreviations for frequency ranges

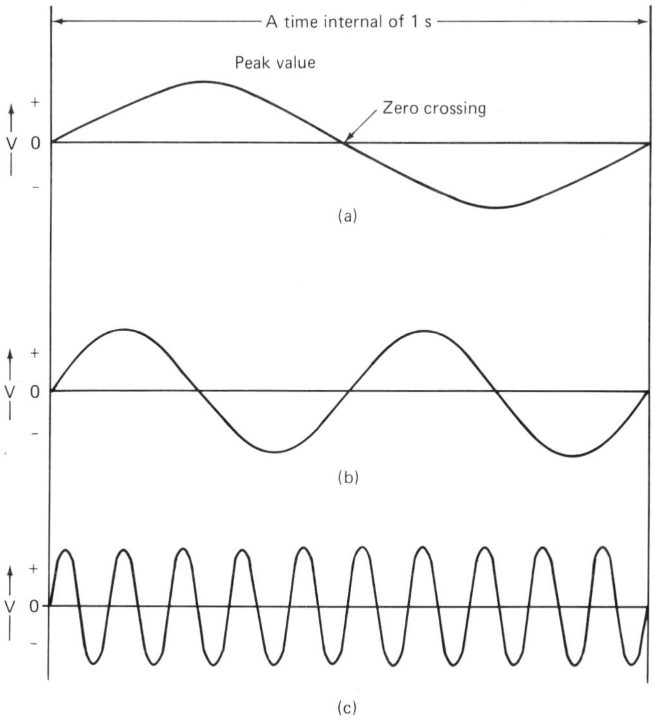

Figure 5.4 Sinusodial voltages of various frequencies. (*a*) 1 Hz: 1 cycle per s, (*b*) 2 Hz: 2 cycles per s, and (*c*) 10 Hz: 10 cycles per s.

Frequency	Classification	Abbreviation
10 to 30 kHz	Very low frequencies	vlf
30 to 300 kHz	Low frequencies	lf
300 to 3000 kHz	Medium frequencies	mf
3 to 30 MHz	High frequencies	hf
30 to 300 MHz	Very high frequencies	vhf
300 to 3000 MHz	Ultra high frequencies	uhf
3000 to 30000 MHz (3 to 30 GHz)	Super high frequencies	shf

Figure 5.5 The frequency spectrum.

are shown in Figure 5.5. Notice that very low frequencies include the audible range. Medium frequencies are typically used for AM radio broadcasting. Very high frequencies are used in television and FM radio transmitters. Super high frequencies are used for radar which locates aircraft, ships, and other objects.

This frequency spectrum continues to increase. The various colors perceived by our eyes are extremely high-frequency sine waves that are no longer radio waves—they become visible light.

Because sine waves are always changing sign (and even spend short moments at 0 v), a question about usefulness of sine wave voltages to heat a room may arise. You only want to pay for the heating effect of a voltage, because this provides some value to you.

Mathematical analysis shows that taking the root mean square of a sine wave voltage provides a value of voltage that represents the "heating effect" of the sine wave voltage. This "root of the mean squared" voltage is abbreviated as rms. A sine wave voltage with an root mean square value of 1.5 v produces the same heat output in a resistor that a dc voltage of 1.5 v does, as shown in Figure 5.6.

There are many ways to measure the voltage of a sine wave, as shown in Figure 5.7. The most obvious is to measure the peak voltage, v_{peak}, the highest voltage that the sine wave achieves. Mathematical anal-

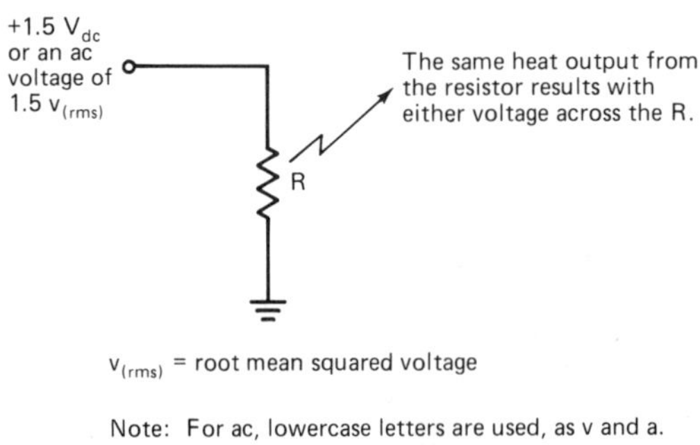

Figure 5.6 The heating effect of a sine-wave voltage.

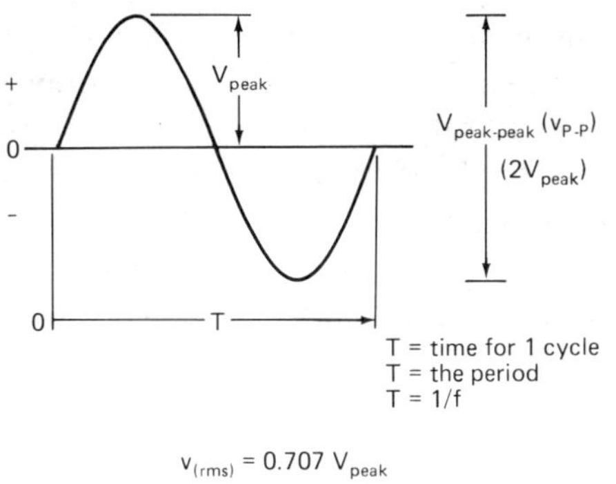

Figure 5.7 The voltage amplitude of a sine wave.

ysis shows that the root mean square voltage is 0.707 of this peak voltage value. Also notice in this figure that the time T taken to complete one cycle is called the period and is measured in seconds.

There is sometimes confusion with the time axis of a sine wave because both time units (seconds) and phase units (degrees) are used, as shown in Figure 5.8. Note that one cycle is completed in 360 degrees (360°). When two sine waves of the same frequency are compared in their relative positioning or timing, phase is usually used. So you will hear of being "in phase" (both sine waves going up and down exactly in step) or "180° out of phase" (when one sine wave is at its positive peak, the other sine wave is at its negative peak). All possibilities in

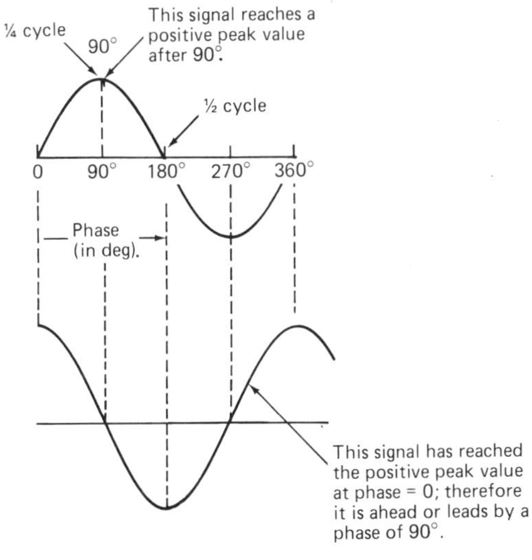

Figure 5.8 The timing of a sine wave.

between can exist. For example, Figure 5.8 shows that the lower sine wave is already at its positive peak when the phase angle of the upper sine wave is 0°. The upper sine wave will take 90° in phase to reach its positive peak, so the lower sine wave is ahead of the upper sine wave by 90° in phase, or it leads the upper sine wave by a phase of 90°. Phase shifts typically exist between two sine waves of the same frequency, so you may hear some of these strange sounding ways to indicate this shift in the relative timing between two sine waves.

Oscilloscopes

A useful instrument that allows observation of sine waves, and other waveforms, is the oscilloscope. The display screen of an oscilloscope is shown in Figure 5.9. An oscilloscope can be thought of as a small TV set that displays voltage waveforms. In the oscilloscope display shown in Figure 5.9*a*, there are no voltages present; therefore, all that is displayed is a horizontal line across the viewing screen of the oscilloscope.

If an oscilloscope were used to look at the voltage that is provided by an ac wall outlet, the 60 Hz sine wave (supplied by the electric power company) would be displayed, as shown in Figure 5.9*b*. To complete 60 cycles in 1 s means that one cycle of this sine wave has to be completed in one-sixtieth of a second or 16.67 milliseconds (ms).

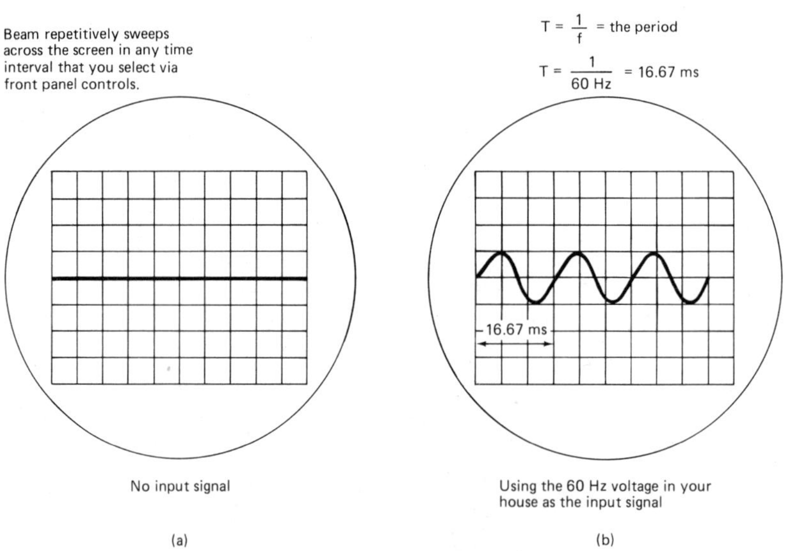

Figure 5.9 The oscilloscope, a TV screen that displays voltages. (*a*) No voltages present and (*b*) seeing a sinusoidal voltage.

Electric utility companies use 60 Hz because the human eye cannot perceive any change in illumination when incandescent bulbs are powered with 60 Hz. If a significantly lower frequency were used, the light output of these bulbs would flicker and create an unacceptable visual disturbance.

An oscilloscope can be used to observe many types of different voltage waveforms. For example, Figure 5.10*a* is a display of the voltage that is provided by a battery. In this case, there is simply an upwardly displaced constant amplitude line, indicating that the battery is providing a dc voltage.

A square wave is shown in Figure 5.10*b*. Many of the basic time bases or "clocks" in digital systems make use of square waves.

A sawtooth waveform is shown in Figure 5.10*c*. This waveform linearly ramps up to a maximum positive voltage, then nearly instantly resets to zero, and then starts ramping up again. This particular waveform is called a "sawtooth" waveform because it resembles the teeth

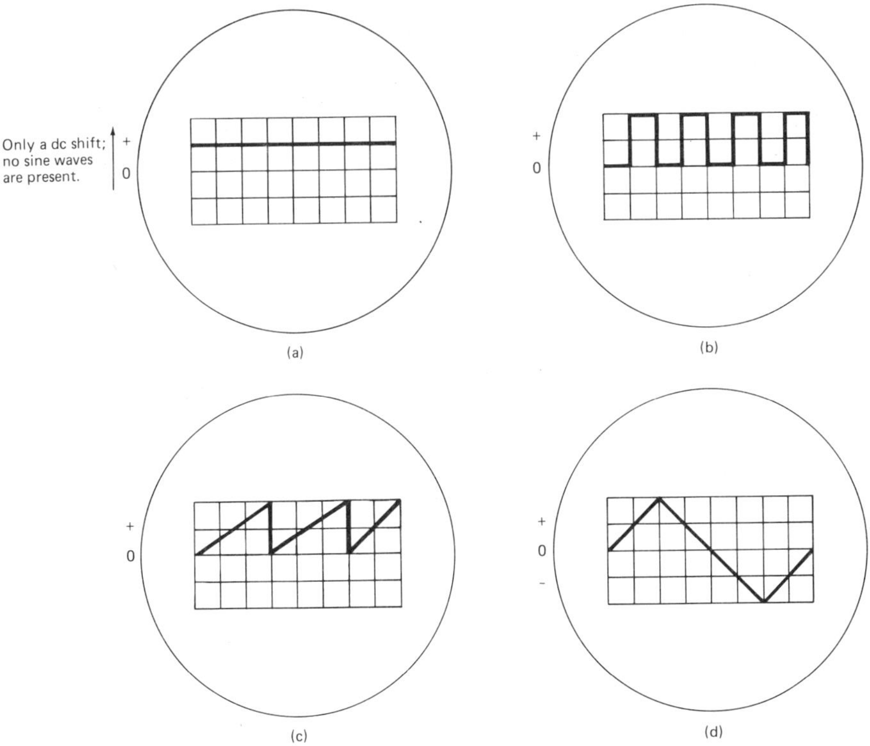

Figure 5.10 Oscilloscope displays of some typical voltage waveforms used in electronics. (*a*) Looking at the voltage of a battery, (*b*) a square wave, (*c*) a sawtooth waveform, and (*d*) a triangle waveform.

on a carpenter's saw. Sawtooth waveforms are often used in linear systems to provide "voltage sweep circuits," voltages that are the analog of time.

Figure 5.10*d* shows an interesting waveform that linearly increases from zero to a maximum positive value and then changes and starts linearly decreasing in value, goes through zero and reaches a maximum negative value, and then starts increasing again. This is called a "triangle" waveform.

There are many different possible waveforms, and the oscilloscope is necessary in order to examine the details of them. Oscilloscopes are therefore very useful in the design of electronic circuits and systems and also are useful in troubleshooting and repair.

Amplifiers

An example of common electronic systems that make use of sine waves is hi-fi amplifiers. One of these systems is diagrammed in Figure 5.11. Note that there is an input signal source, like an AM tuner, an FM tuner, a tape deck, a record player, or a microphone. This input signal source provides a complex voltage waveform that results from the many simultaneous frequencies that are present in music or speech. These individual frequency components vary in amplitude and phase depending on the content of the programming material.

This low-level input signal is fed to an amplifier to increase the magnitude of the voltages and to modify the tonal response. For example, the base (low frequencies) can be boosted, the high frequencies can be attenuated, an equalizer can be added (to correct for the acoustical problems of a room), or in other ways the tonal "shape" of a musical selection can be custom tailored to suit a listener.

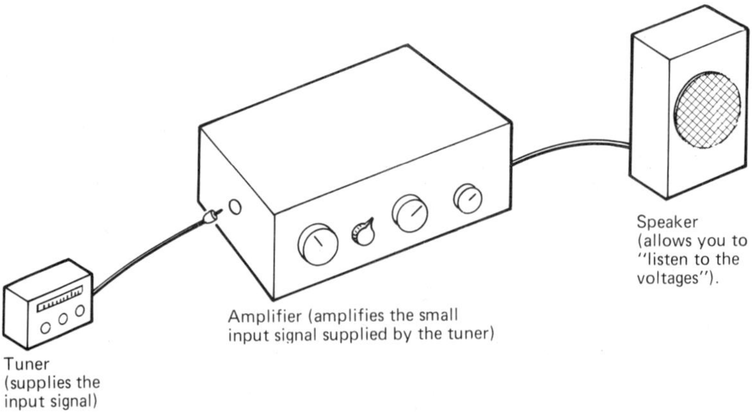

Figure 5.11 A pictorial layout of a hi-fi audio system.

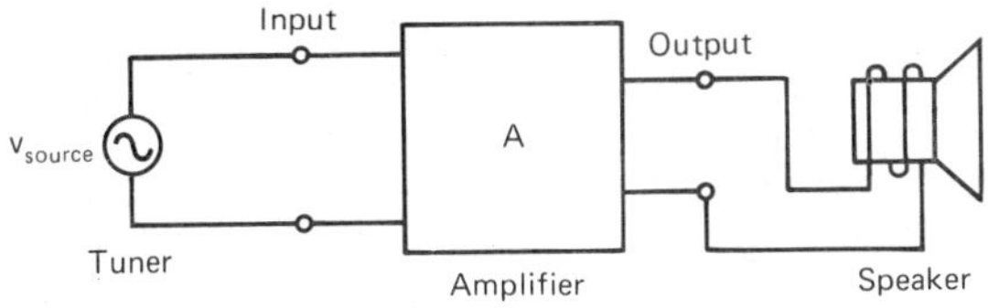

Figure 5.12 A basic amplifier.

After coming out of the amplifier, the signal then goes to a loudspeaker. This is the transducer that converts the varying electrical voltages into sound: fluctuating pressure waves in air that our ears respond to.

This electronic amplifying system can be diagramed as shown in Figure 5.12. The input signal source is simply represented by the symbol for a sine wave voltage source. This signal enters the input of the amplifier, A, and the output of this amplifier is connected to the loudspeaker. We will now take a look at amplifiers, because providing amplification for low-level signals is a major function of linear systems.

In general, amplifiers have a pair of input terminals and a pair of output terminals, as shown in Figure 5.13*a*. If the terminals associated with the input (terminals 1 and 2) and the terminals associated with the output (terminals 3 and 4) are completely separate and electrically isolated from each other, this is called a "four-terminal amplifier."

Many amplifiers make use of only three terminals, as shown in Figure 5.13*b*. Notice that terminal 3 is common to both the input and the output. This is the case when a transistor is used as an amplifier, because a transistor has only three leads. With two leads of a transistor associated with the input signal and two leads associated with the output, one of the leads of the transistor must be common to both the input and the output, as shown in Figure 5.14. This gives rise to the

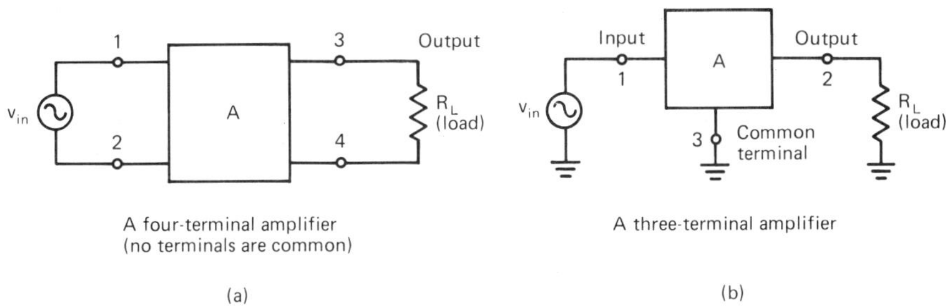

Figure 5.13 Four-terminal and three-terminal amplifiers. (*a*) Some amplifiers have two leads for the input and two separate leads for the output, and (*b*) many amplifiers have a common ground lead and therefore use only three terminals.

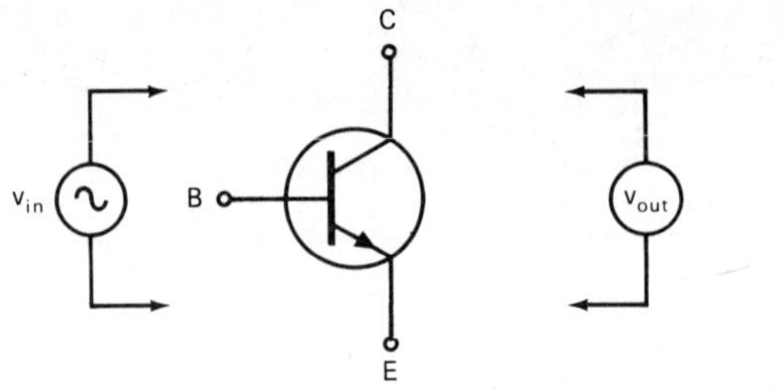

Figure 5.14 Applying an input voltage to, and obtaining an output voltage from, a transistor.

three basic ways a transistor can be connected as an amplifier, shown in Figure 5.15: the common-emitter, the common-collector, and the common-base amplifiers.

As mentioned earlier, the common collector (also called the "emitter follower") has a voltage gain of at most 1, but the common-emitter and the common-base amplifiers can have voltage gains larger than 1 (special circuitry and special transistors can be used to obtain voltage gains of up to about 4000 for a single transistor common-emitter amplifier as was done in the LM3900 IC product).

The way that these three basic transistor amplifiers are biased is shown in Figure 5.16. Notice where the input signal is applied to the transistor and also where the output signal voltage is obtained in each

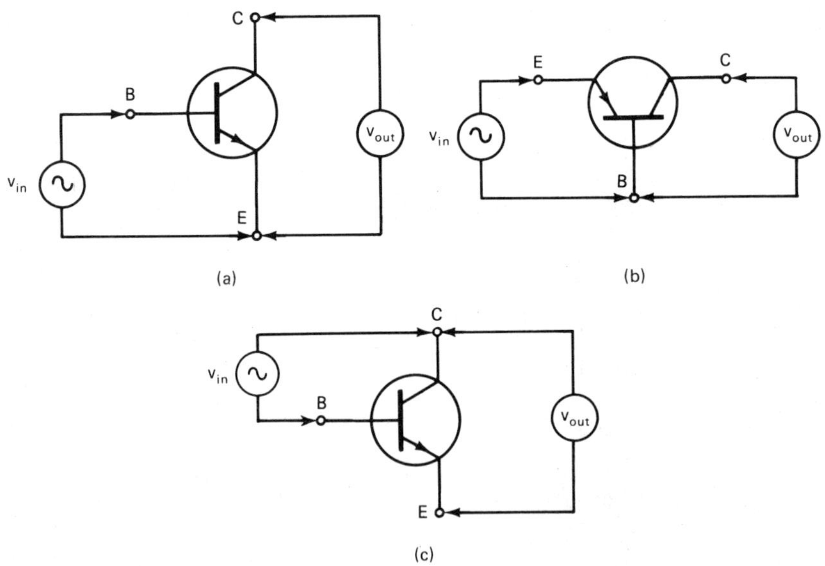

Figure 5.15 The three possible basic transistor amplifier connections. (*a*) Common emitter, (*b*) common base, and (*c*) common collector.

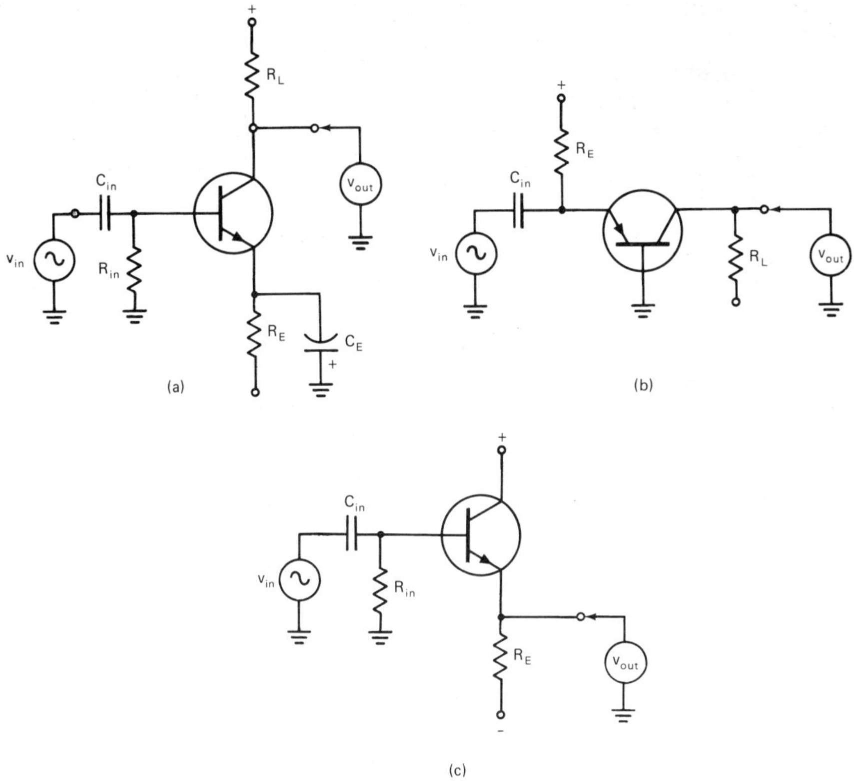

Figure 5.16 Biasing the three basic transistor amplifiers. (*a*) common emitter, (*b*) common base, and (*c*) common collector (the emitter follower).

of these amplifiers. The input coupling capacitors, C_{in}, couple the ac signal from v_{in} but prevent the v_{in} source from disturbing the dc voltage that exists at the transistor input lead. Capacitors block dc but allow ac to pass through, as we will see in the next chapter. These biasing circuits have been simplified by making use of both positive and negative power supply voltages (biasing these amplifiers from a single power supply voltage would add to the complexity of the biasing circuitry).

Notice that the common terminal of all of these amplifiers is tied to an ac ground (that is, there is no ac signal voltage at these common terminals). An ac ground exists at any of the power supply voltage nodes in a circuit (this is the case with the common collector) or at the circuit side of a large-valued capacitor that is tied to ground and is said to be "bypassing that circuit node to ground" (the case for the common emitter) or at an actual connection to ground (the case for the common base). There are many other biasing adaptations of these basic circuits in which dc biasing voltages may be applied to the common terminals, but these types of circuits then use a large-valued capacitor from these common terminals to ground to ensure that there are no ac signals at these common terminals.

The amount of voltage amplification or voltage gain A_v of an amplifier can be expressed as a number, such as 1000, or it can be expressed in decibels (dB) where

$$dB = 20 \log A_v$$

An A_v of 1000 could therefore also be expressed as 60 dB. A listing of the conversion from numeric values of voltage gain to decibel values is given in Figure 5.17. The benefit of using decibels is that the overall gain obtained from a cascade of amplifier stages is simply the sum of the individual decibel gains of each amplifier. If numeric values were used for the individual gains of each amplifier, the overall gain would be given by the product of these numeric gains. This makes decibels easier to work with in the design of electronic systems.

A_v	A_v in dB
0.707	-3
1	0
3	9.5
10	20
30	29.5 (30)
100	40
300	49.5 (50)
1000	60
3000	69.5 (70)
10000	80
30000	89.5 (90)
100000	100

Figure 5.17 Voltage gain expressed in decibels.

Classes of amplifiers

Amplifiers are often referred to as being operated in a certain class, such as class A, class AB, class B, class C, or class D. This designation refers to the amount of conduction of the amplifying device, the transistor. If the transistor or transistors that are associated with the output stage of an amplifier are always conducting (for the full 360° of an input sine wave), it is called a class A amplifier. When each transistor separately handles only one-half of the signal (one handles the positive swing and the other handles the negative swing), the conduction of each transistor is only 180° out of the full 360° of an input sine wave, and this is called a class B amplifier. If these two transistors were each biased to carry some quiescent current so that each of the transistors would then conduct slightly more than 180°, a class AB amplifier would result. When the transistors conduct for less than 180° (such as only a very short time interval, once per cycle), a class C amplifier results. Class C amplifiers are generally restricted to applications that use a tank circuit as the load for the amplifying transistor. The "flywheeling" action of the tank circuit provides a continuous sine wave output voltage even though the amplifying transistor only "kicks" or supplies short pulses of current to the tank circuit once each cycle. This is much like the short pushes that can be given, once each cycle, at the appropriate time, to keep a child's swing going. Class C amplifiers are usually restricted to high frequencies, but they have application whenever a tank circuit is used as the load for the amplifying transistor. Finally, a class D amplifier is a switching amplifier that supplies a varying burst of current pulses that have to be converted into an output sine wave voltage by a filter network. There is much interest in switching-type circuits because the transistor is operated in an ON/OFF mode and can provide very high efficiencies. In class D amplifiers, there are problems with the high frequency switching "noise" that is generated and also with the quality of the resulting output signal.

"Biasing" a transistor to amplify

The diagram of a very simple three-terminal amplifier, a common-emitter transistor amplifier circuit, is shown in Figure 5.18. The current source between the collector and the emitter terminals simulates the transistor. There are two components of current that are supplied by this current source. One is a dc value of current, the biasing current, that results because of the dc input voltage. "Biasing" a transistor simply means providing the proper operating dc voltages and dc currents for a transistor so that the transistor will exhibit gain (it would certainly be strange, and we would be getting something for nothing, if a transistor could exhibit gain all by itself, without requiring any external sources of energy—this would be nice, although theoretically

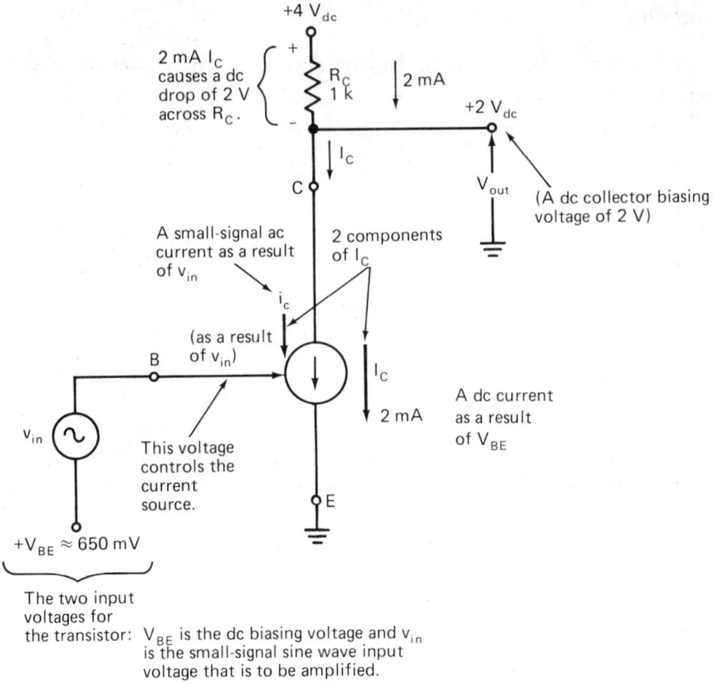

Figure 5.18 Partially turning on a transistor so that it will amplify an ac signal.

impossible, because batteries or other sources of electrical energy would not be needed in electronic equipment). The emitter dc biasing current is 2 milliamps (mA), and the collector dc biasing voltage is 2 V.

Another component of the current of this current source is a small signal variation, an ac current, i_c. This ac current results because of the ac input signal voltage. So, in order to make an ac amplifier out of a transistor, two different input voltages have to be supplied to the transistor: a dc biasing voltage and an ac signal voltage.

In the collector lead of the amplifying transistor, there is a 1-kilohm (kΩ) collector load resistor R_C which is connected to the positive 4-V power supply. Therefore, the 2-mA dc emitter-biasing current that flows through this 1-kΩ resistor creates a voltage drop of 2 V across this resistor. Figure 5.19 shows that if there is no ac input signal, the output voltage will simply sit at 2 V, the "quiescent" collector voltage (quiescent means quiet and is the condition when no signals are applied). This collector voltage is biased at one-half of the 4-V power supply voltage in order to allow a maximum ac signal swing at the collector.

The biasing problem in a common-emitter transistor amplifier can be represented as shown in Figure 5.20. A biasing circuit is needed

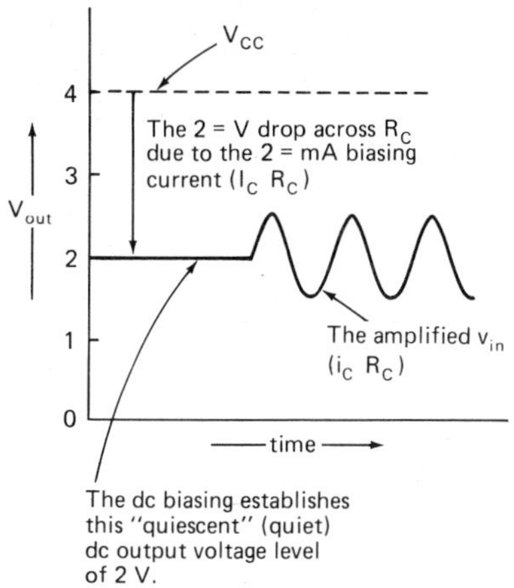

Figure 5.19 Displaying the output voltage waveform.

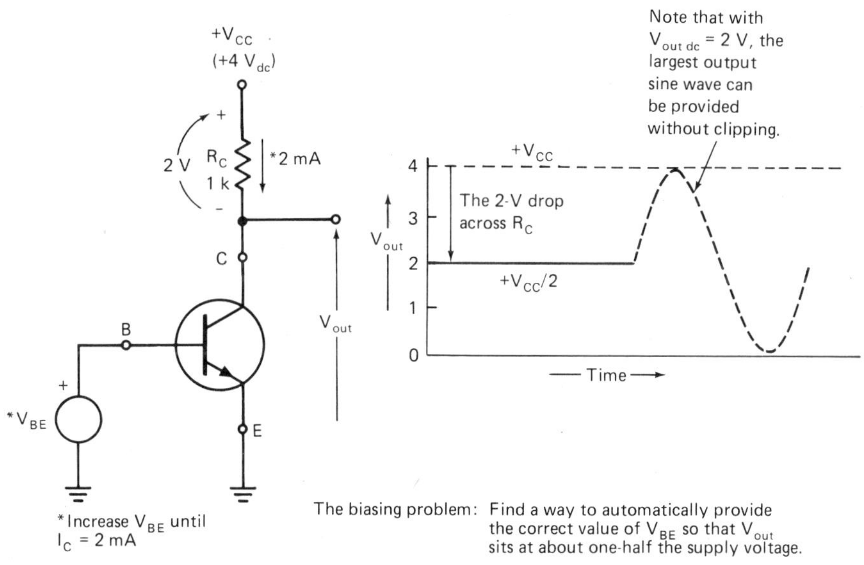

Figure 5.20 Biasing a transistorized ac amplifier.

that will automatically provide the proper dc base-emitter voltage that is required to cause the dc collector current to have a value of 2 mA. Because of the use of a 1-kΩ collector load resistor, this gives the desired quiescent dc output voltage level of 2 V. This is shown with the dotted-line waveform that moves up 2 V and down 2 V (4 V peak to peak) from the 2-V quiescent level.

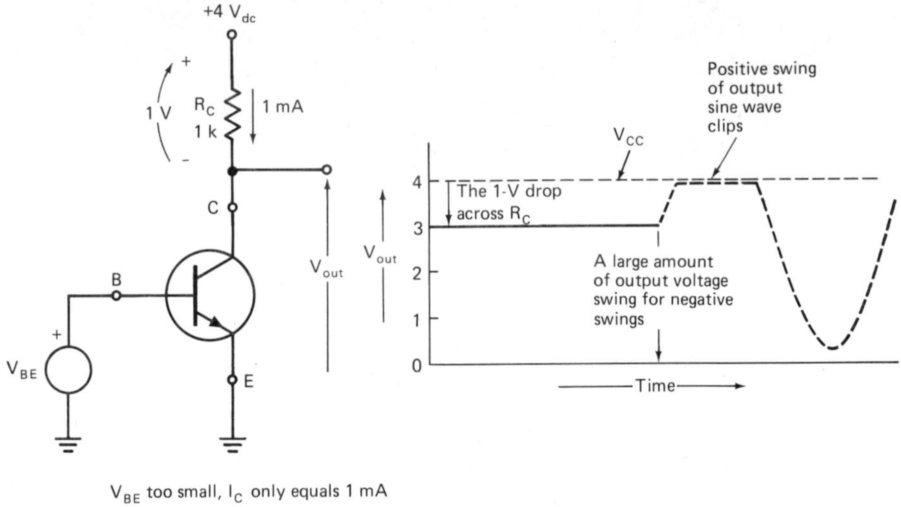

Figure 5.21 Large-signal positive-peak clipping.

As an example of an improper bias, assume (as a result of using a biasing circuit with poor control) that the base-emitter voltage is too small, and, as a result, the collector current is only 1 mA. In this case there would be a voltage drop of only 1 V across the 1-kΩ collector load resistor, and the quiescent dc collector voltage would therefore move up to 3 V. This collector voltage can now only move up 1 V before the positive swing of the output voltage is clipped (a very bad distortion) at 4 V, as shown in Figure 5.21. Notice that a larger 3-V negative swing can be obtained. This collector-biasing voltage is not centered, and this limits the maximum output signal that can be provided before getting positive peak clipping. Completely biasing a transistor either ON or OFF is used in digital circuits and provides the abrupt ON/OFF type of digital signals such as positive-going pulses, negative-going pulses, and square waves.

Clipping a voltage waveform is a very typical distortion that is most likely to occur when the volume is turned up too high in a hi-fi system. At high-output power levels, something is eventually going to overload and then cause distortion. If the loudspeakers can take the higher power level, generally the limit will be voltage clipping within the circuits of the amplifier. When clipping occurs, the output doesn't sound good; it is no longer the faithful reproduction of the input signal that is implied by "high fidelity."

An opposite problem in biasing a transistor amplifier is shown in Figure 5.22, in which instead of having too small a dc base-emitter voltage, this voltage is too large (again, as a result of a bad biasing circuit design). In this case, 3 mA of collector current results. The 1-

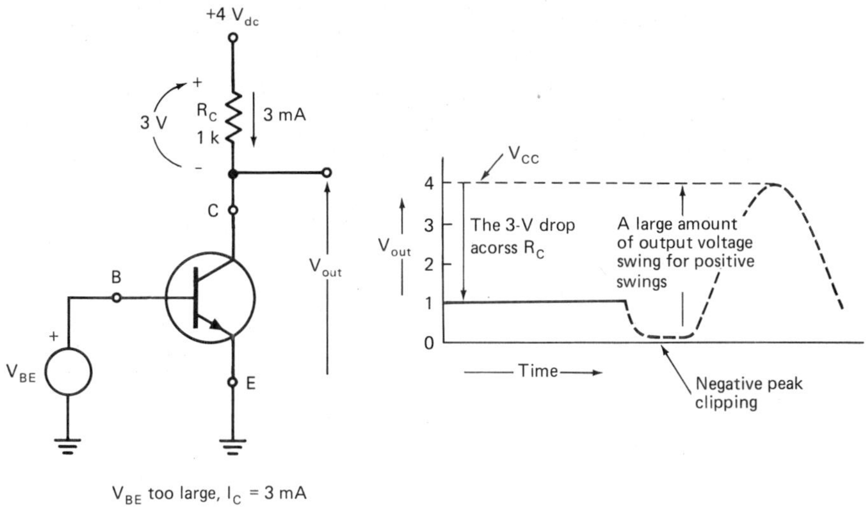

Figure 5.22 Large-signal negative-peak clipping.

kΩ collector load resistor therefore creates a 3-V drop, and the quiescent level at the output is now only 1 V. In this case, the collector voltage can swing 3 V positive but can swing only 1 V negative before negative peak clipping occurs. Clipping on either side of a voltage waveform is undesirable because it represents distortion and because in either case the output signal is no longer a faithful reproduction of the input signal.

An elaborate transistor biasing circuit is shown in Figure 5.23. This

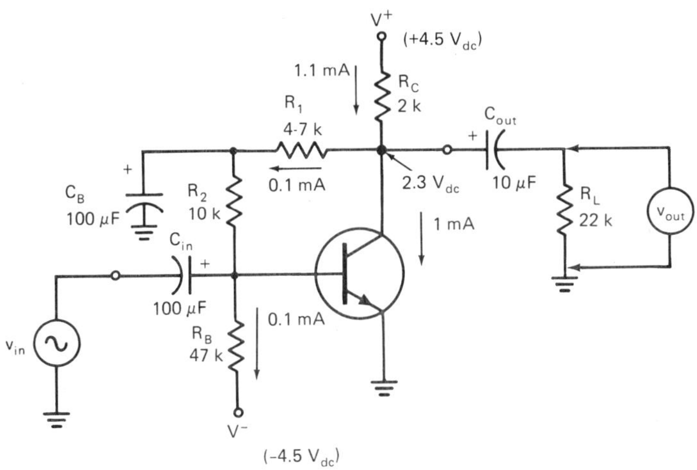

Figure 5.23 Voltage feedback biasing.

circuit makes use of the biasing resistor R_B to pull a dc current of approximately 0.1 mA through resistors R_1 and R_2. The top end of R_B will sit at a dc voltage of approximately 0.5 V (the base-emitter voltage of the transistor). Therefore there is approximately 5 V across this 47-kΩ resistor, which establishes the 0.1 mA biasing current flow. This biasing current is supplied by the collector load resistor R_C, along with the 1 mA of biasing current for the collector of the transistor. This biasing circuit operates by making use of the fact that the voltage drop that is created across both of the resistors R_1 and R_2 by the flow of the 0.1 mA biasing current will exist between the base and collector of the transistor. This fixes the lower end of the collector load resistor R_C at a predetermined voltage and thereby determines the current flow through this resistor. This current flow, which is in excess of the 0.1 mA biasing current, must then flow into the collector of the transistor (the circuit automatically adjusts itself so that this occurs). If the base current of the transistor is kept small in comparison to the biasing current flow through R_B, a very predictable biasing circuit results in which the operation depends mainly on the resistor values used and is essentially independent of the characteristics of the transistor (V_{BE} and the current gain, β).

The object of biasing circuits is to maintain the dc biasing conditions on the transistor essentially independent of the parameters of the transistor (that are subject to variation from transistor to transistor, such as β). The proper biasing of individual transistors has been a rather difficult design problem and has required the use of quite a few resistors and capacitors for each of the transistors within an amplifier circuit. Today, many circuit designers use an IC op amp to accomplish a function that has been handled in the past with only a single transistor. In many cases, the cost of the op amp is less or at least comparable to the cost of a single transistor and the performance of the op amp circuit is usually greatly improved (this is especially true for dc and low-frequency applications, but discrete transistors are still used to provide very good high-frequency amplifiers). The op amp is the subject of Chapter 7, and it will then become clear why the op amp has been such a popular building block for linear systems designers.

The emitter follower

A very useful amplifier is the emitter follower (also called a "common-collector" amplifier). The emitter follower has a voltage gain of slightly less than 1, so emitter followers are not used to amplify signals. They "buffer" or isolate a load from a signal source because an emitter follower has a high input resistance and therefore will not load the input signal source. An emitter follower also has a low output resistance and this allows it to supply large load currents.

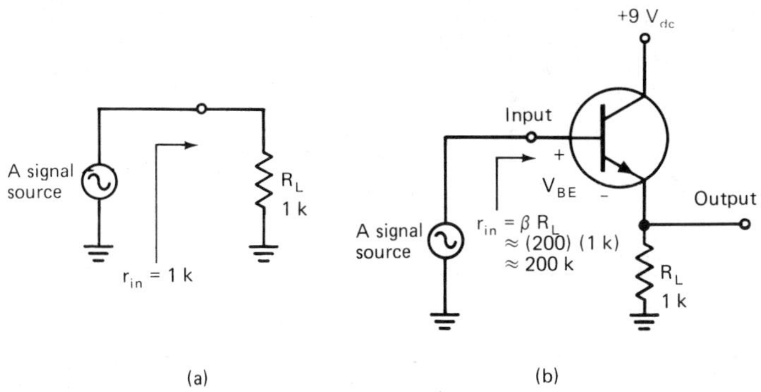

Figure 5.24 The very useful emitter follower. (*a*) A load resistor "loads" the signal source, and (*b*) an emitter follower isolates the load.

As shown in Figure 5.24*a*, we are assuming that the 1-kΩ load is too small a resistance value (an excessive load) for the signal source to drive. When an emitter follower is used, as shown in Figure 5.24*b*, and the output of this emitter follower is connected to the 1-kΩ load, a large input resistance is obtained (at the base of this transistor) and the signal source will then not be loaded. The signal source will "see" an input resistance that is beta (β) times the load resistor, or 200 kΩ (for $\beta = 200$). The emitter follower is used as an isolating amplifier and not only prevents input loading, but it also provides (or sources) output current to the load resistor.

Figure 5.25 shows how to bias an ac-coupled emitter follower. Two resistors provide a voltage divider across the 9-V power supply to bias the base of the NPN transistor at 2.9 V. The emitter resistor, the 1-kΩ resistor from the emitter to ground, establishes a quiescent biasing current flow of 2.3 mA in the transistor. A capacitor C_{in} is used to couple the ac input signal so that the dc biasing of the base will not be disturbed. Also an output capacitor C_{out} couples the output signal to the external 1-kΩ load. Both of these capacitors prevent interference with the dc biasing of the transistor. When the emitter follower is used, the signal source will not be loaded by the 1-kΩ load resistor because of the isolation that is provided.

Notice that the emitter follower can provide, or source, large currents to the load resistor, but it can only absorb, or sink, the current that is supplied by the emitter-biasing resistor. This is one of the shortcomings of the emitter follower. A much better isolation amplifier is the voltage follower that makes use of an op amp (discussed in Chapter 7).

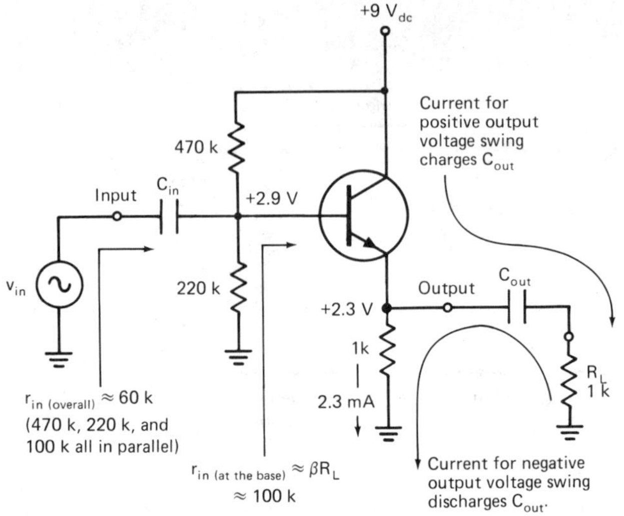

Figure 5.25 An ac-coupled emitter follower.

A power amplifier

The schematic diagram of a transistorized power amplifier that you can hook up using the components that are available in the lab kit is shown in Figure 5.26. This type of circuit is often called a "public address system," or a "PA system," and also represents one-half of an intercom or a paging system. The output voltage of the microphone (the crystal earpiece is used as a microphone) is amplified and eventually provides a high-level signal that can operate the loudspeaker.

On the input side of the diagram, a PNP transistor (terminals 73, 74, and 75) is used as an emitter follower to provide a high input impedance so the microphone will not be loaded and it will therefore provide a larger input signal voltage. The other PNP transistor in the lab kit is operating as a common-emitter amplifier and is using a transformer in its collector circuitry. This is a class A amplifier. This transformer couples the signal into the input of the power amplifier output stage.

The power amplifier output stage is made with the two NPN transistors. Two silicon diodes are used to provide a dc voltage bias that is supplied to terminal 46 of the input transformer. This dc voltage is therefore applied to the bases of both of the NPN transistors. The 100-Ω resistor from the common-emitter connection to ground establishes a quiescent current flow of 3 mA in each of these NPN transistors (this therefore makes this a class AB output stage). The output transformer provides a way to couple the low-impedance (10 Ω) loudspeaker to the collectors of the power amplifier.

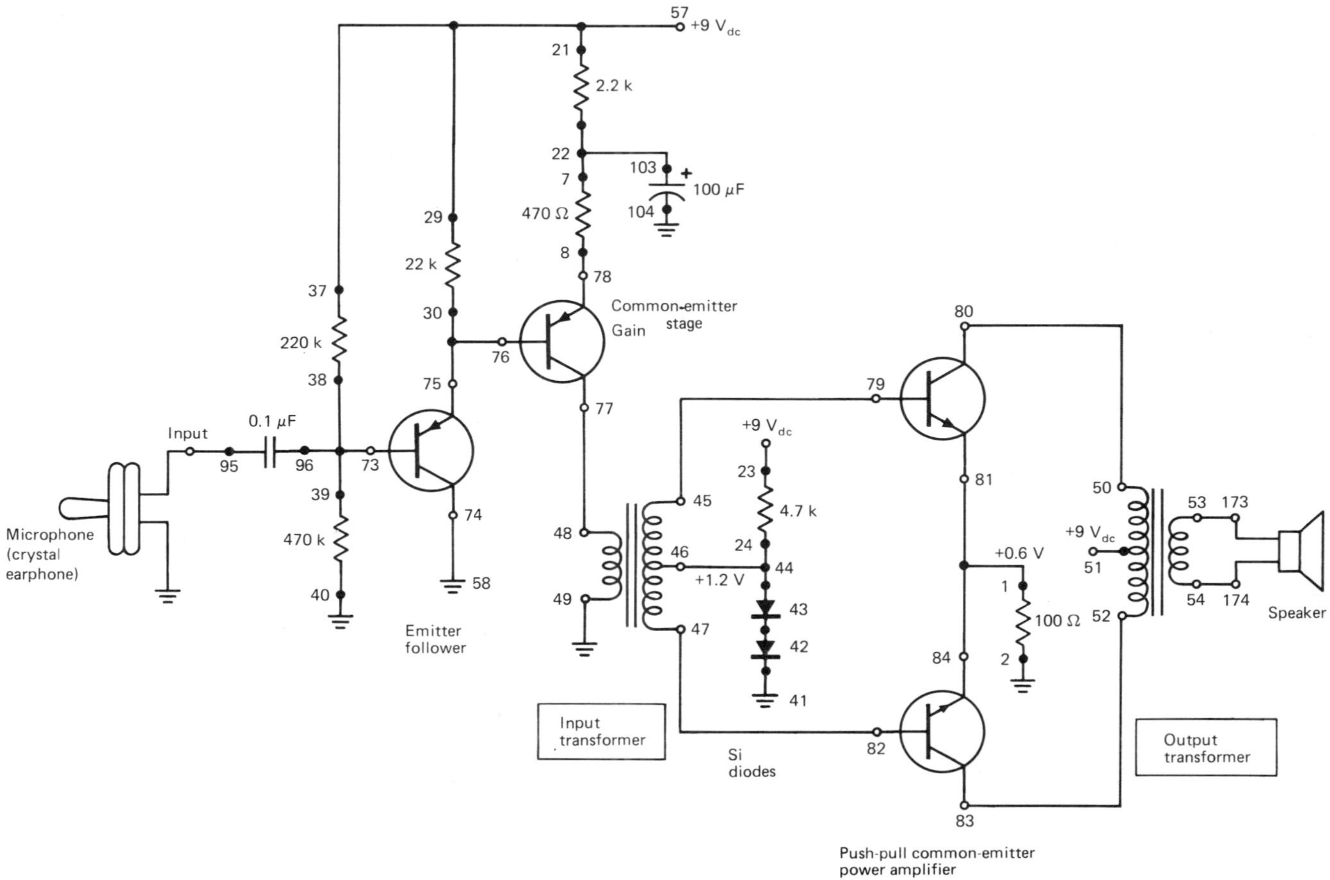

Figure 5.26 A transistorized audio power amplifier.

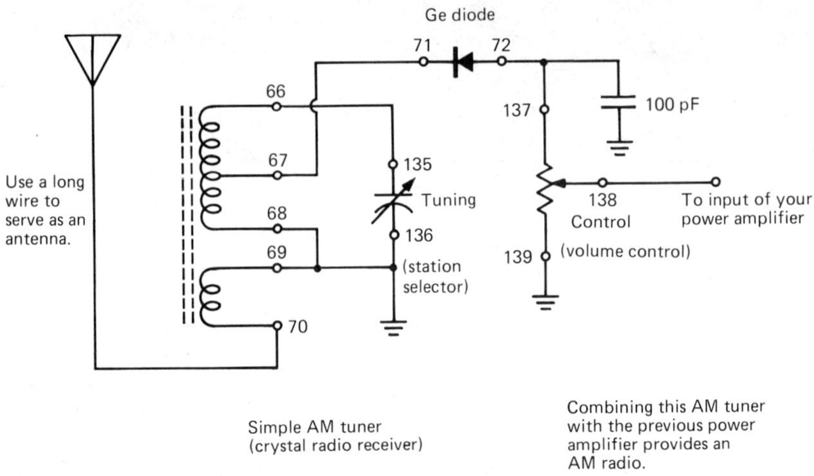

Figure 5.27 An AM tuner for the power amplifier.

This is a useful amplifier. You can talk into the microphone and hear your voice coming out of the loudspeaker. But, if you get the microphone too close to the loudspeaker, audio coupling will make a squealing sound because the resulting acoustical feedback from the speaker back into the microphone converts this amplifier into an oscillator (this squealing, or oscillation, caused by pickup of the speaker output by the microphone, is a problem with all public address systems).

A simple AM tuner can be used as the signal source for this power amplifier, as shown in the schematic of Figure 5.27. A long wire from terminal 70 of the antenna coil provides the receiving antenna that picks up the signals that are sent out by AM radio transmitting stations. The tuning capacitor (available from terminals 135 and 136) provides station selection. A germanium diode is used as a simple AM "detector" (a rectifier that will conduct only in one direction—we will have more to say about rectifiers later in this chapter). The Control is used to adjust the volume.

By tying the wiper of this control to the input of the previous power amplifier, a relatively simple AM radio reciever results. If you live far away from an AM radio station, you might have to add a better antenna (a longer wire) to pick up a signal, and the ground point of the tuner may require a connection to a cold-water pipe or to a metal rod driven into the earth.

Measuring AC Voltages

When ac circuits are studied, it becomes necessary to measure ac voltages. Figure 5.28 shows how to convert the digital multimeter into an

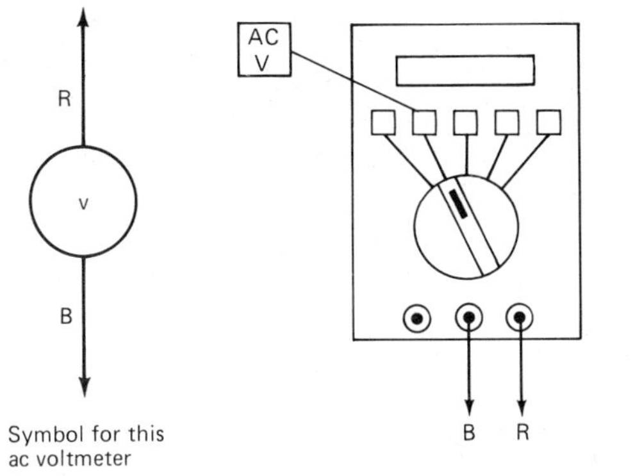

Figure 5.28 Setting up the digital multimeter to measure ac volts.

ac voltmeter that indicates the root mean square value of a sinusoidial waveform. The Selector switch is simply positioned to the lower-right segment (the red area that is labeled "ac V"), and the test leads are plugged in as if measuring dc volts. The symbol for an ac voltmeter uses the lowercase *v*, as opposed to an uppercase *V* (which signifies a dc voltage). Always connect the black probe to ground and connect the red lead to the voltage to be measured.

Some interesting experiments that demonstrate the problem of the ever-present undesired power line hum pickup can be demonstrated with this ac meter, as shown in Figure 5.29. Simply plug the leads into the meter, turn the meter on (the lowest voltage scale 2-V full scale will result), and lay the leads widely spaced apart across your desk. The high input resistance of this meter will probably provide an ac voltage reading of something close to 500 mV.

Try connecting the test leads of the meter across the terminals of various resistors in the lab kit and notice that the hum pickup prob-

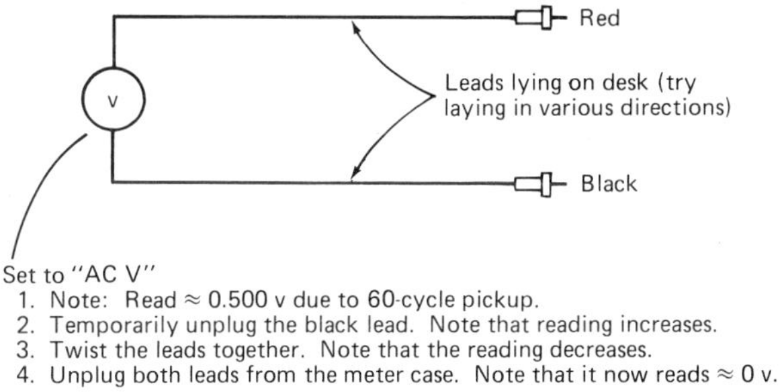

Figure 5.29 Measuring power-line hum pickup.

lem is worst with the highest resistance possible: leaving these leads open circuited. This is the 60-cycle power line pickup that is present because of the wiring that exists in the walls and ceilings of most buildings. Notice that if you temporarily unplug the black lead from the meter case, the meter reading will increase because the presence of the black lead was tending to cancel some of the signal that was picked up by the red lead. If the same hum voltages simultaneously exist on both meter leads, the hum will cancel. You can achieve this cancellation by twisting the two meter leads together. Even though the meter leads are still left open at the ends, you will notice a decrease in the ac voltmeter reading when the leads are twisted together. This trick of twisting wires together to reduce undesired signal pickup is used in most electronic systems.

A final experiment is to simply unplug both of the meter leads from the meter body and notice that the voltmeter now reads essentially 0 V (because there are no long leads exposed to pickup hum). The problems of hum pickup can be easily understood with this series of experiments with this ac meter.

The basic generation of a voltage can also be demonstrated, as shown in Figure 5.30. The input transformer is used to function simply as a coil of wire. If a permanent magnet (one of the small magnets that hold notes to your refrigerator door will work) is rapidly moved back and forth very close to the top of the input transformer, a reading will be obtained on the ac voltmeter if this meter is connected to terminals 45 and 47 of the input transformer. This demonstration shows that three things are needed to generate a voltage:

1. A conductor (the wire that is making up the winding between terminals 45 and 47 of the input transformer)
2. A magnetic field (the magnetic field, in this case, provided by the permanent magnet)

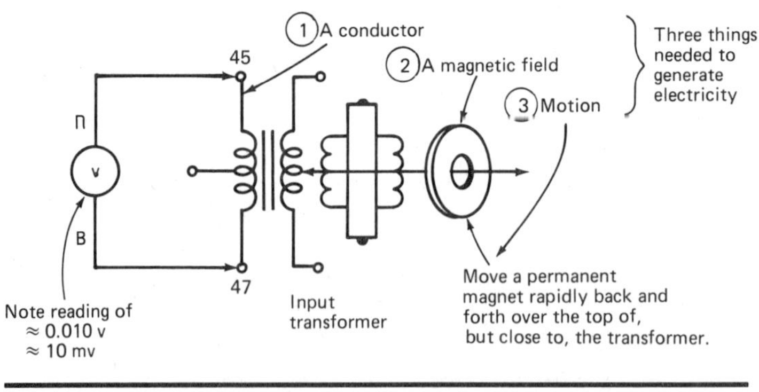

Figure 5.30 Generating electricity.

3. Motion (relative movement between the wire and the magnetic field)

There must be relative motion between the wire and the magnetic field. In this case, the magnetic field is moved and the wire is held stationary. These three basic things, a conductor, a magnetic field, and motion, are necessary on all forms of generators or alternators, whether they are the simple "generators" (actually alternators) that power bicycle lights or the giant alternators that are used in hydroelectric dams.

Building a Useful AC Test Oscillator

In order to do some experiments with ac voltages, an ac signal source is needed. This will be provided by a test oscillator that is made out of the components that are available in the lab kit, as shown in Figure 5.31. The NPN transistor (available from terminals 79, 80, and 81) is used in an oscillator circuit. An additional NPN transistor (available on terminals 82, 83, and 84) is used as an output emitter follower.

The output transformer is used to feed a signal from the collector circuit (the output) of the oscillator back into the base circuit (the input). It is this regeneration, or positive feedback, that creates the oscillation. An input signal is amplified and is then fed back in a reinforcing way so that a constant amplitude sine wave is provided at the output of this test oscillator circuit. The Control is used to vary the amount of the signal that is fed back. This will be used to adjust the magnitude of the ac output voltage.

To provide a useful test oscillator, a pure, single-frequency sine wave is needed. An undistorted sine wave is obtained by slowly increasing the Control from zero, and, using the ac meter to measure the output voltage provided at the emitter (terminal 84), the Control is adjusted to provide an ac output voltage of approximately 50 mv (0.05 v). This low output voltage level is necessary so that this oscillator circuit will stabilize and provide a good output sine wave without regenerating and producing a square wave. The frequency of this test oscillator is very close to 500 Hz.

The symbol for this test oscillator is the same symbol that is used for a sine wave voltage source, and it is shown in the lower right-hand corner of Figure 5.31. The additional information that this ac voltage is 50 mv in amplitude and that the frequency is 500 Hz is included within parenthesis next to the symbol. This test oscillator will be used as a signal source in many measurements and demonstrations of ac circuitry, so don't take it apart right away.

This 500 Hz oscillation can be converted into an audible signal so you can hear it by making use of either the loudspeaker or the crystal earpiece. If the low-resistance (10-Ω) loudspeaker is connected directly across the output terminals (53 and 54) of the output transformer, the

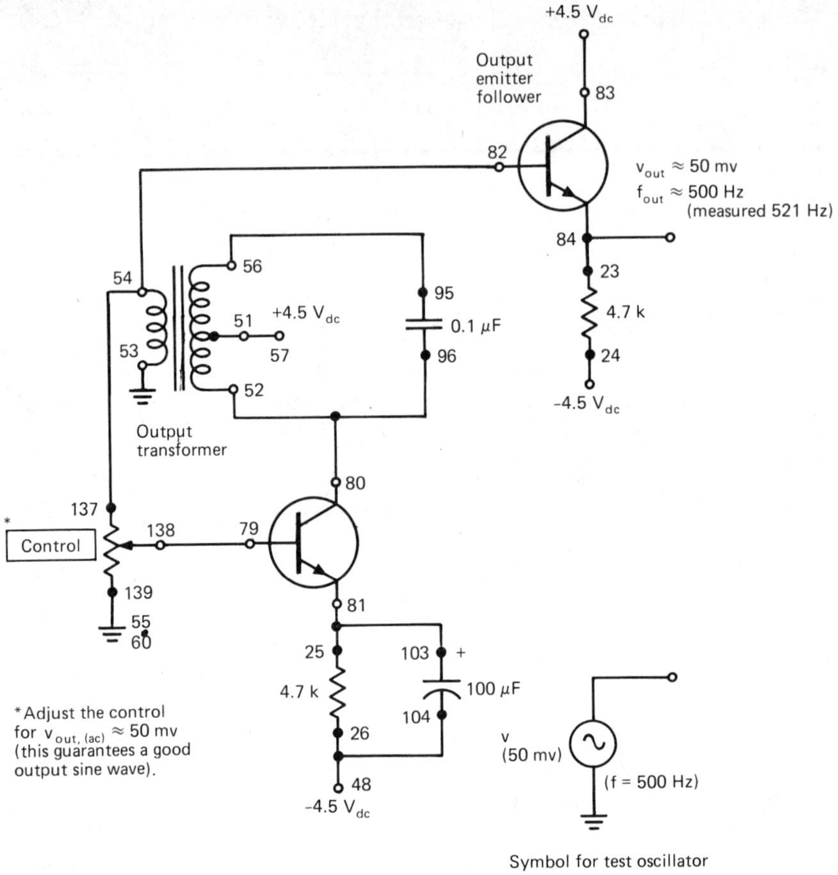

Figure 5.31 A useful 500-Hz test oscillator.

oscillation will stop. The oscillator circuit is loaded too heavily, which reduces the voltage gain, and therefore the oscillation stops. Instead of using the loudspeaker, the high-impedance crystal earpiece will be used and this can be connected directly across the large winding of the output transformer, terminals 50 and 52, as shown in Figure 5.32. This allows you to hear the 500 Hz tone being produced by this test oscillator without any significant loading on the circuit.

Adding a common-emitter amplifier

One of the available PNP transistors is used to make a common-emitter amplifier for the output of the test oscillator, as shown in Figure 5.33. Split power supplies (both +4.5 V and −4.5 V) are used, so it is relatively easy to bias this transistor. The emitter current is supplied from

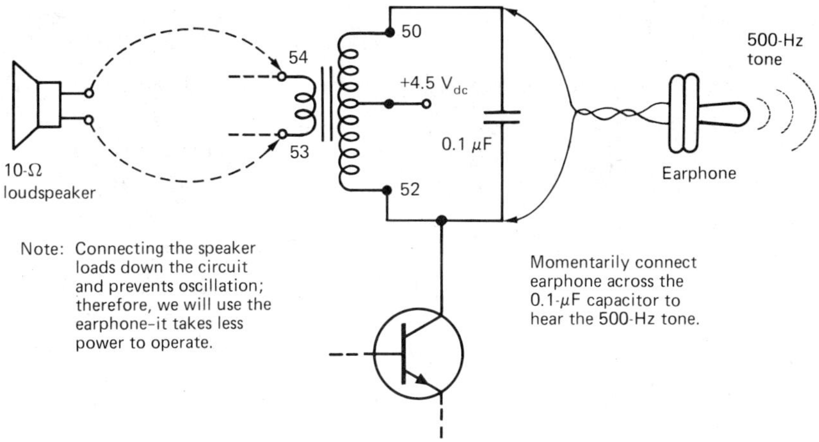

Figure 5.32 Listening to the test oscillator.

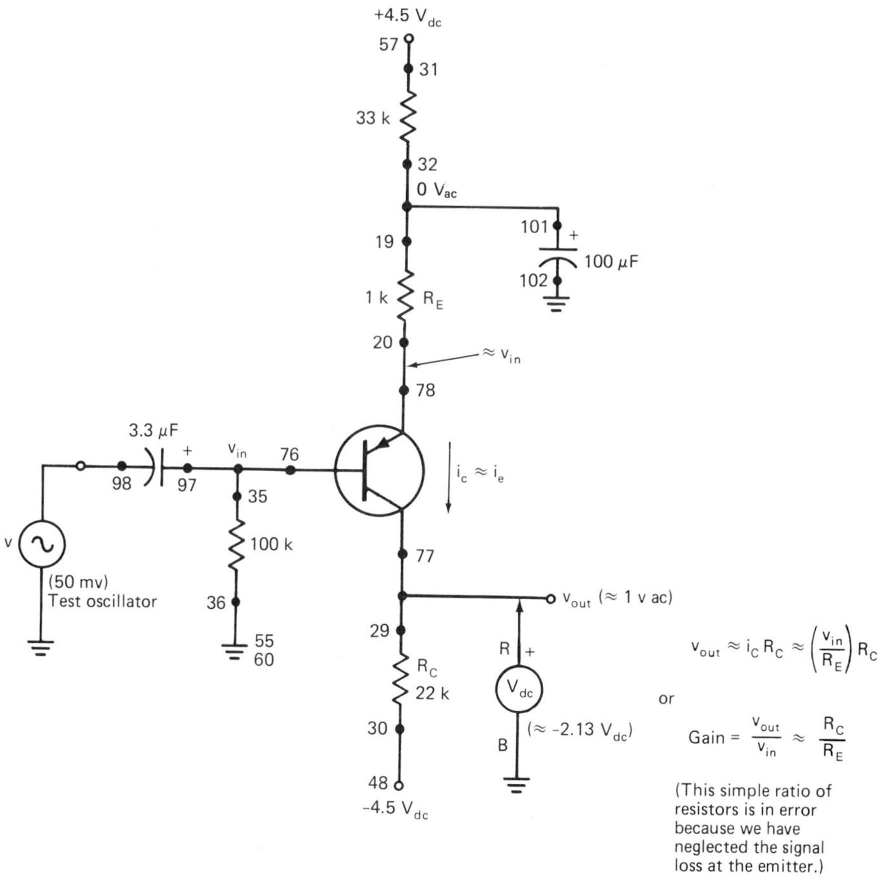

Figure 5.33 Adding a common-emitter amplifier to the test oscillator.

the positive supply voltage through the sum of the 33- and the 1-kΩ resistors, a total resistance of 34 kΩ. This emitter current essentially passes through to the collector, and the smaller collector resistor of 22 kΩ provides a dc output voltage of −2 V. This collector-biasing voltage allows room for an adequate ac voltage swing on the output of this amplifier. The gain of this amplifier stage can be estimated as approximately the value of the collector resistor, 22 kΩ, divided by the value of the emitter resistor, 1 kΩ, or 22. This is sufficient gain to amplify the low-level 50-mv signal to provide nearly a 1-v signal at the output of this amplifier.

Measuring the voltage gain of the common-emitter amplifier

The ac voltage gain of the added common-emitter amplifier can be measured, as shown in Figure 5.34. The ac input signal directly at the base of the amplifying PNP transistor is approximately 46 mv, and the voltage at the collector is approximately 855 mv. The ratio of this output voltage to the input voltage is the ac voltage gain. This ratio is 855 to 46, or a voltage gain of 19. (This is smaller than the previous estimate for the gain of this stage because the transistor adds a component to the total value of the emitter resistance.)

This is an inverting amplifier (but this is not obvious when using the meter to measure voltage gain). This means that as the input voltage swings up, the output voltage does the opposite: it swings down. These voltages are 180° out of phase. In most audio applications it doesn't matter if the signal is inverted, although signal polarity can be important in many other applications.

The 100-microfarad (μF) capacitor prevents any ac signals from existing at the top end of the 1-kΩ resistor R_E. As far as ac signals are concerned, the top end of this resistor is operating at ac ground (this can be verified by measuring, with the ac voltmeter, that no ac voltage exists at this point). Therefore, for an ac signal, the emitter resistance is 1 kΩ; but, for dc biasing, the emitter resistance is the sum of the 33- and the 1-kΩ resistors. The 100-μF capacitor is used as an emitter bypass capacitor, a very important application for capacitors.

The signal of the test oscillator has been elevated to almost the level of 1 v with the addition of the common-emitter, PNP, single-transistor amplifier. While this large amplitude test signal is available, we will take a look at some special circuits that require large ac signals.

A Look at Some Rectifier Circuits

A large application area for diodes is within power supplies, where they are used to convert from an ac voltage to a dc voltage that will serve

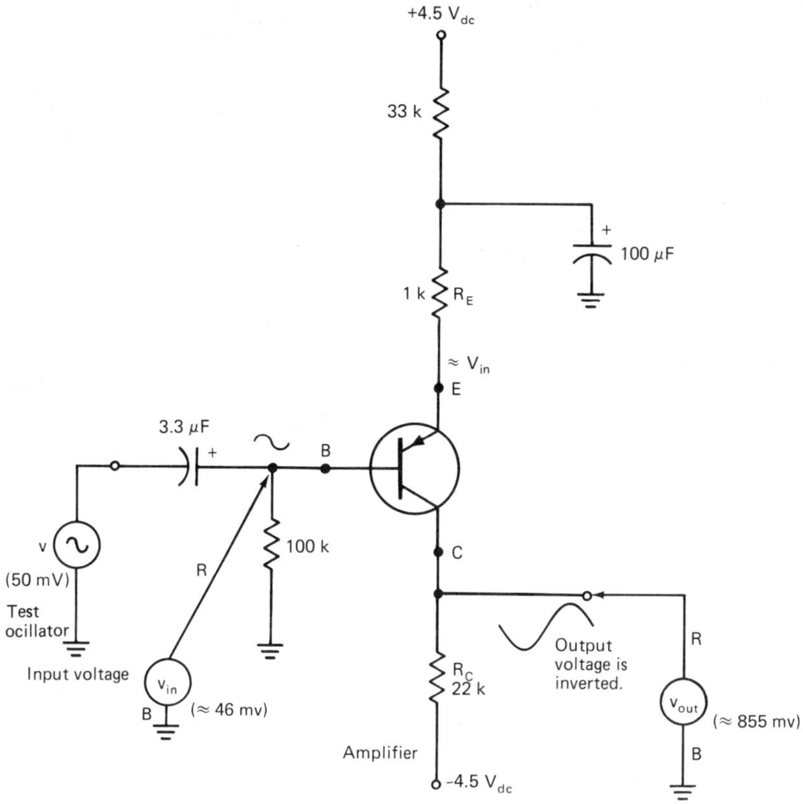

$$A_v = \frac{-V_{out}}{V_{in}} \approx \frac{-855 \text{ mv}}{46 \text{ mv}} \approx -19$$

$$\text{Note: } A_v \approx \frac{-R_C}{R_E} \approx \frac{-22 \text{ k}}{1 \text{ k}} \approx -22$$

Figure 5.34 Measuring the ac voltage gain of the common-emitter amplifier.

as the unregulated input for a power supply. In the previous example of the AM tuner (that was mentioned in this chapter and was used to provide an input signal to the power amplifier), a germanium diode was also used as a rectifier. In this application, this rectifier is more properly known as an AM detector. The word "rectifier" is generally used for the diodes that are used to convert ac voltages into dc voltages. Rectifier action can now be demonstrated by making use of the amplified ac signal that is available from the test oscillator.

As shown in Figure 5.35, an output coupling capacitor of 10 μF is used to couple the amplified ac signal from the test oscillator to a 220-kΩ load resistor. The ac voltage that exists across this 220-kΩ resistor now swings both positive and negative with respect to ground.

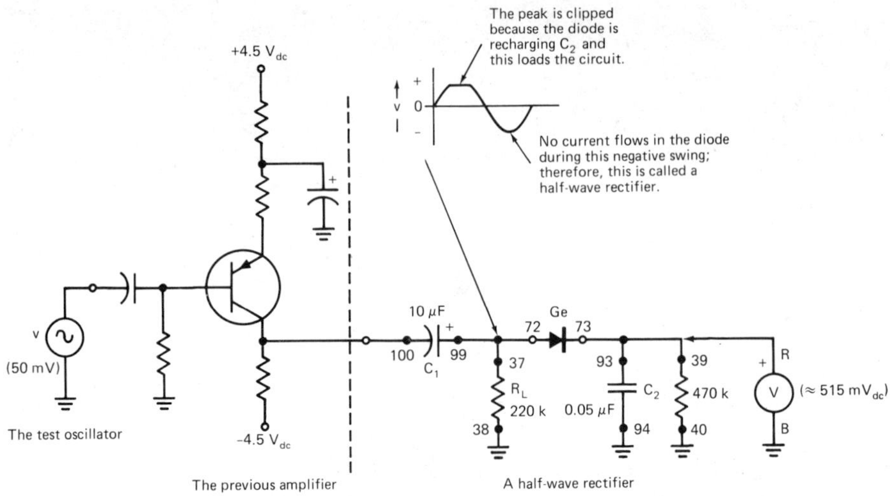

Figure 5.35 A half-wave rectifier.

The germanium diode will conduct during the positive excursion of this waveform, and it won't conduct during the negative excursion. As a result, the voltage across the final 470-kΩ resistor will be a positive dc voltage. The dc voltmeter will indicate a value of about 515 mV because the 0.05-μF capacitor will charge to a voltage that is one diode-forward-voltage-drop less than the peak value of the ac input signal. This capacitor is recharged during every positive cycle of the 500-Hz test oscillator ac waveform, and it supplies the dc discharge current that flows through the 470-kΩ resistor.

Following a positive peak in the ac input voltage, the voltage across the capacitor starts decreasing because of the current loss to the 470-kΩ resistor. It will soon happen that the voltage across the capacitor is larger than the instantaneous magnitude of the now decreasing ac input voltage, because the capacitor is still "remembering," or storing, nearly the peak value. Therefore, the diode simply opens up or disconnects. This diode stays off until very nearly the next positive peak of the ac input voltage. So for almost the total time interval of the ac input sine wave, the diode remains nonconducting. The capacitor provides a charge "reservoir" during the longer time interval that exists while the diode remains off and supplies the dc current to the load. This is a typical example of a half-wave rectifier. Only a very small part of the positive half of the sine wave passes through the diode.

Figure 5.36 shows the changes needed to make a full-wave rectifier. Full-wave rectifiers are often used in power supplies that operate from the 110-v ac power that is available from a wall outlet. This high voltage is first stepped down to a lower voltage with a step-down trans-

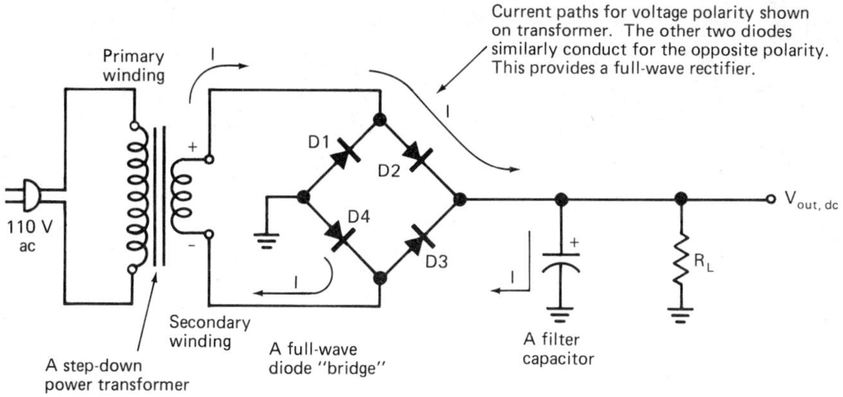

Figure 5.36 A full-wave rectifier.

former (this function of a transformer will be described in Chapter 6). Then, a particular connection of four diodes, called a full-wave diode bridge, is used for the rectifier. Each pair of these diodes alternatively conducts on each half of the ac input voltage.

When the sine wave on the upper terminal of the secondary of the transformer (shown in Figure 5.36) is positive, current flows out of the positive lead of the transformer, through diode D_2, through the capacitor (a charging current), back through the ground connection, and then through diode D_4 and back into the negative terminal of the transformer. Diodes D_1 and D_3 are reverse biased during this time interval and therefore do not conduct.

When the sine wave voltage polarity changes and the lower side of the secondary winding becomes positive, diode D_3 will forward bias into the capacitor and diode D_1 will return the current into the (now negative) upper lead of the secondary winding. In this way, two of the diodes conduct on alternate half cycles and thereby provide a pulse of charging current into the capacitor on every half cycle. Compared to a half-wave rectifier, this full-wave rectifier reduces the amount of ripple (the amount of voltage change) that exists across the filter capacitor, because this capacitor is now charged twice per cycle instead of only once per cycle as exists with a half-wave rectifier circuit.

The dc output voltage from either of these rectifiers (half wave or full wave) is usually not constant enough nor predictable enough to be used to supply voltage directly to electronic circuitry. A voltage regulator has to be added. A very simple shunt voltage regulator was discussed in Chapter 4 that can be connected (through a series-dropping resistor) to the unregulated output voltage that is provided by either one of these rectifier circuits to make a complete power supply for electronic circuits. Special integrated circuits, called "series voltage regulators" (which will be considered in Chapter 7), are also available that supply only the current required from the output of the regulator and are therefore more efficient than a simple shunt regulator.

To understand how a half-wave rectifier can function as a detector for an AM radio receiver, we have to consider what is involved in providing the signal that is transmitted by an AM radio station, as shown in Figure 5.37. The "carrier frequency" for an AM radio transmitter is a radio-frequency sine wave signal with a frequency somewhere within the range that has been allocated for the AM radio band: 550 kHz to 1.5 MHz. This carrier frequency establishes the transmitting frequency of the radio station and is needed to "carry" the program material across space to a radio receiver. After this transmitted signal is picked up by a radio receiver, the carrier has done its job (it has "carried" the information to the receiver), and it is no longer needed.

In an AM radio, the programming information consists of audio frequency sine waves. A "modulator" circuit at the radio station is used to constantly change the instantaneous amplitude of the carrier voltage in response to the much-lower-frequency audio programming voltage. The output of an AM transmitter is the amplitude-modulated (AM) carrier shown in Figure 5.37*b*.

The very weak (1 to 100 μv) received signal voltage that is picked up by an antenna is amplified; low-cost AM radios omit this radio frequency (RF) amplifier. Then, all incoming RF carrier frequencies are translated to the same lower frequency, called an "intermediate frequency," or "IF" (intermediate between the RF and the audio frequencies). This is done so that the IF amplifiers can be optimized. Within the receiver, all radio stations now have the same frequency, the IF (this is the key idea of the long popular, high-performance superhetrodyne radio receiver).

The information-bearing audio frequencies are then stripped away from the IF with an AM demodulator, which is also called a detector. For AM radio, the half-wave diode rectifier, shown in Figure 5.38, is

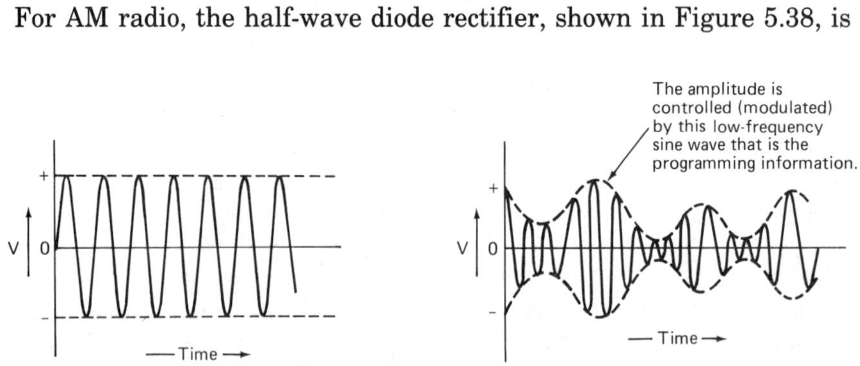

Figure 5.37 The signal transmitted by an AM radio station. (*a*) A constant amplitude and (*b*) an amplitude modulated radio wave.

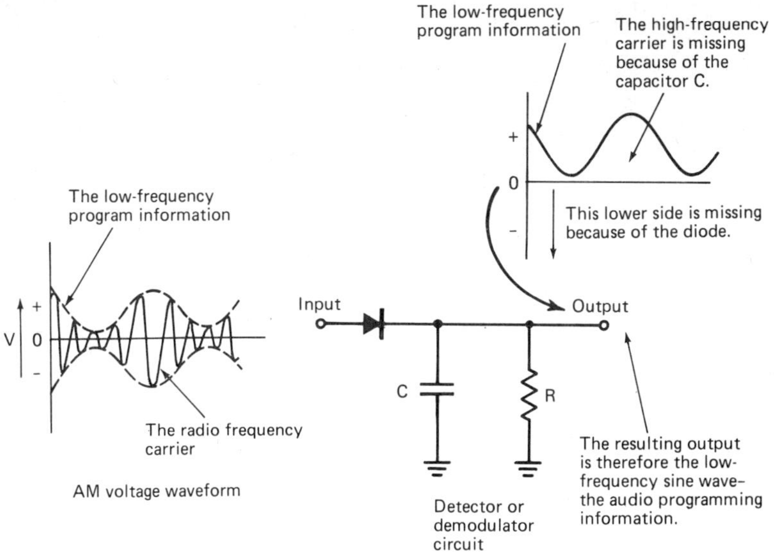

Figure 5.38 Demodulating an AM signal.

often used. The IF signal is ac; it swings both positive and negative. The diode only passes the positive side of this signal; this is the half-wave rectifier action. The capacitor, *C*, then bypasses the high-frequency IF signal to ground. The only thing that is left is the much-lower-frequency audio programming information. This recovered audio signal is then amplified by an audio frequency amplifier and is then passed to a loudspeaker so you can "listen to the radio."

Now that we have more familiarity with ac voltages, we will look at the complications that are introduced when we consider circuit components operating with these ac voltages.

Chapter

6

Circuit Components Operating with Alternating Current

AC circuits are more difficult to analyze than the much simpler dc circuits. Both are included so you will have a better idea of the separate dc and ac operations that simultaneously take place within electronic circuits.

In dc circuits, we found that resistors limit current flow because of their resistance. In ac circuits, inductors and capacitors similarly limit the ac current flow, but this is now because of their inductive reactance X_L or capacitive reactance X_C. Reactance is also measured in ohms, but the value of reactance depends on the frequency f of the ac excitation. For example,

$$X_L = 2\pi fL$$

and

$$X_C = 1/2\pi fC$$

where $\pi = 3.14$
L = the value of inductance, in henrys (H)
C = the value of capacitance, in farads (F)

A "passive" (this means that there are no transistors) ac circuit can consist of R's, C's, and L's. Another parameter, impedance Z (also measured in ohms), is used for the general combination of all of these passive components. For example, the impedance of the series RLC circuit shown in Figure 6.1 can be calculated as

$$Z = \sqrt{R^2 + (X_L - X_C)^2}$$

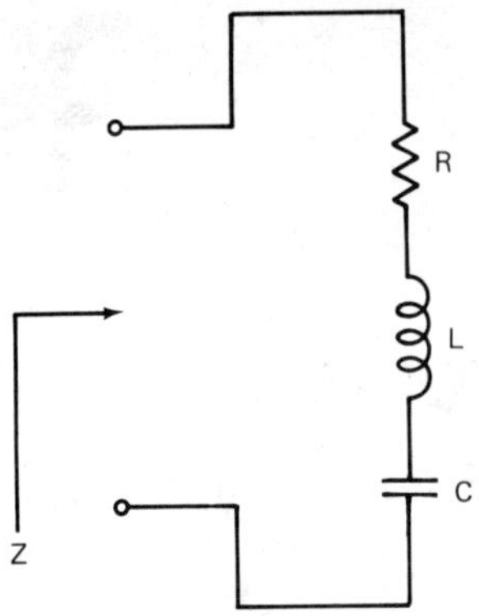

Figure 6.1 The impedance of a series *RLC* circuit.

and any one of these terms can be omitted if that particular component is not used. Consider, for example, the interesting case of $R = 0$ and $X_L = X_C$. The impedance of this series-connected L and C has the very interesting value of zero! (We will consider this series "resonance" condition later in the chapter.)

Resistors

The resistor is the only component that operates in essentially the same way with ac as with dc. There is a "skin effect" at very high frequencies in which the current is restricted to flow in only the outer surface of the material of the resistor. This raises the resistance value because the inner regions of the resistor are not conducting current.

A second nonideal effect exists at lower frequencies as a result of the physical construction of a resistor. An unavoidable stray capacitance appears across the ends of a resistor. The effect of this end-to-end stray capacitance is shown in Figure 6.2, in which the value of the resulting impedance is plotted as a function of frequency. Notice that at high frequencies the impedance falls because of this parasitic capacitance C_s that is shunting the resistor element. The next section will discuss the reason why capacitors more easily allow high-frequency currents to flow; but, for now, be aware of the fact that resistors can have smaller values of impedance than expected at high frequencies. This problem is worse for larger-valued resistors (greater than 10 kΩ) and not much of a problem at all for a 100-Ω resistor. For example, with a 500-kΩ resistor, this stray capacitance starts reducing the impedance value at frequencies of 100 kHz.

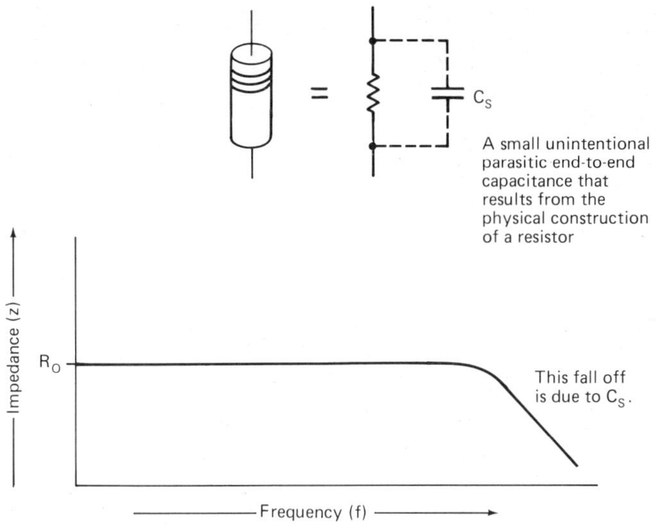

Figure 6.2 Resistance is essentially independent of frequency.

The phase relationship between *v* and *i* in a resistor

In ac circuits, an ac voltage is applied across the components and, as a result, an ac current will flow through the components. The relationship between voltage v and current i within a resistor when connected to an ac voltage source is shown in Figure 6.3 When the voltage is zero, the current is also zero. As the ac voltage reaches a positive maximum value, the current simultaneously reaches a positive maximum value. The current waveform therefore goes up and down exactly in step with the voltage waveform; it goes through zero and reaches the maximum values at the same times as the driving voltage waveform. Therefore, the current flow through a resistor is "in phase" with the voltage across the resistor (this will not be the case with capacitors or inductors). A resistor operating with an ac excitation voltage is not very different from a resistor operating with a dc voltage excitation, especially at low frequencies. Ohm's law for dc circuits can be used for ac circuits when only resistors are involved.

Capacitors

Good capacitors don't pass dc current, so a good capacitor that is connected to a dc voltage source (where the voltage does not change) simply charges to this voltage value and then no more current flows. But, what happens if a capacitor is connected to an ac voltage source?

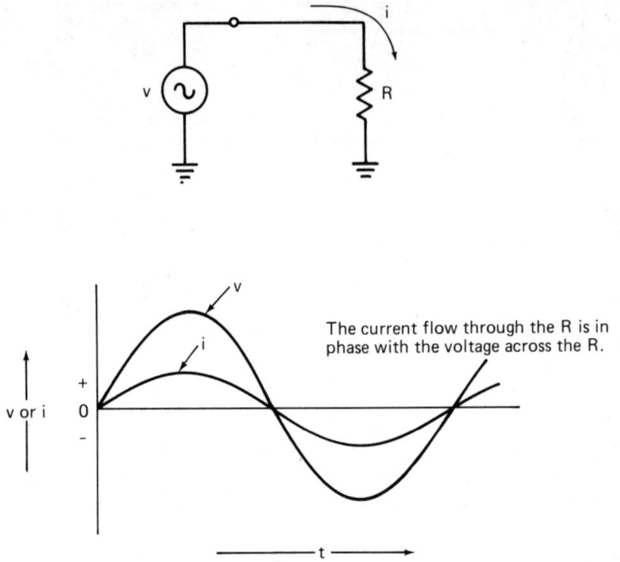

Figure 6.3 The phase relationship between current and voltage in a resistor.

We would like to consider connecting a capacitance of 0.01 μF to the test oscillator and then calculating how much ac current would flow "through" (actually "into" and then "out of") the capacitor. Before we can do this, we first have to again consider the concept of capacitive reactance.

Capacitive reactance

Capacitors do allow ac currents to flow, but the capacitive reactance limits the amount of the current flow. Capacitive reactance is also measured in ohms. The reactance of a capacitor is designated as X_C, and the formula for capacitive reactance is given (again) in Figure 6.4. Note that X_C is said to vary "inversely" with frequency, because frequency is in the denominator of this expression. When working with high frequencies, the expression for capacitive reactance is divided by the large value for frequency, and this gives small values of capacitive reactance. Capacitive reactance therefore falls as the frequency increases (there will be more about this shortly).

To find out how much current flows if a 0.01-μF capacitor is connected to the test oscillator, the capacitive reactance is first calculated. Using 500 Hz as the value of frequency f and a capacitance value C of 0.01 μF, a capacitive reactance of almost 32 kΩ (Figure 6.4) is calculated. With the 50-mv signal of the test oscillator (the added amplifier

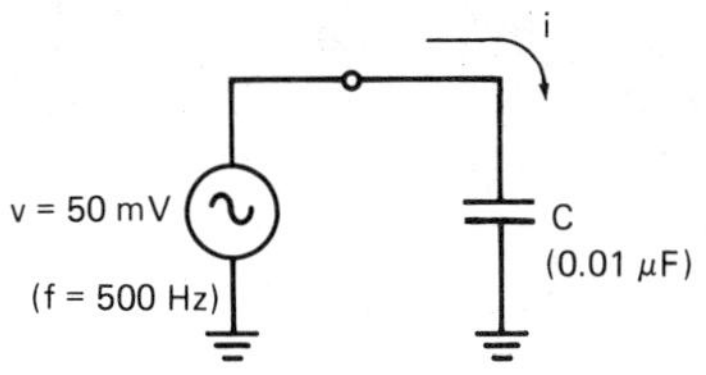

How much i flows?

Capacitors don't have resistance (R)—they have capacitive reactance (X_C).

Where $X_C = \frac{1}{2\pi fC}$ Ω

and $\pi = 3.14$

For example: A 0.01-μF capacitor at 500 Hz has an X_C given by

$$X_C = \frac{1}{2\pi(500)(0.01 \times 10^{-6})}\ \Omega$$

$$X_C = \frac{1}{3.14 \times 10^{-5}}\ \Omega$$

$$X_C = 31.8\ \text{k}\Omega$$

Therefore, with v = 50 mv:

$$i = \frac{v}{X_C} = \frac{50\ \text{mv}}{31.8\ \text{k}\Omega} = \frac{5 \times 10^{-2}\ \text{v}}{3.18 \times 10^{4}\ \Omega}$$

$i = 1.57 \times 10^{-6}$a = 1.57 μa
(Our multimeter doesn't directly measure ac current.)

Figure 6.4 Capacitors have capacitive reactance.

will no longer be used), the ac current flowing in this capacitor is then 50 mV divided by 32 kΩ, or almost 1.6 μA. With such a small value of capacitance, 0.01 μF, not much current flows because the capacitive reactance is so large (at the frequency of only 500 Hz). A larger value of capacitance would have a smaller value of capacitive reactance. For example, a 0.1-μF capacitor would reduce the capacitive reactance by a factor of 10 to 3.2 kΩ.

Figure 6.5 shows, graphically, that as the frequency is increased, the capacitive reactance decreases. Increasing the frequency by a factor of

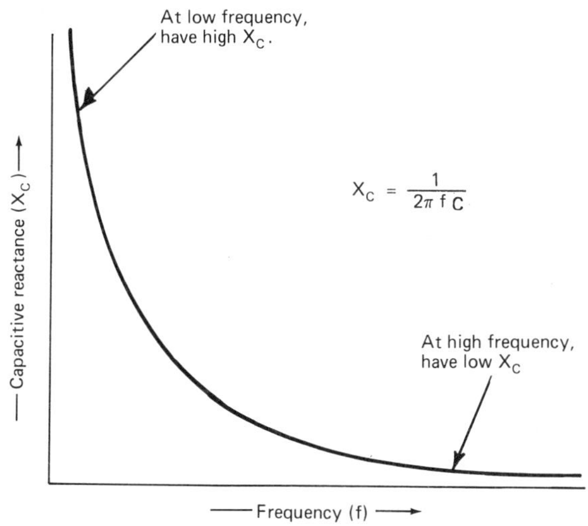

Figure 6.5 Capacitive reactance varies with frequency.

10 will therefore reduce the capacitive reactance by a factor of 10. Keep in mind that capacitors have reduced reactance at higher frequencies.

The phase relationship between *v* and *i* in a capacitor

The phase relationship between the ac current *i* that flows in a capacitor and the ac voltage *v* across the capacitor is the strange result shown in Figure 6.6. As the voltage increases from zero (point A on the waveform), the current waveform is seen to just have moved through a positive peak and to then start decreasing. The current waveform is ahead of the voltage waveform, and therefore *current leads voltage in a capacitor*. In the ideal capacitor shown here (with no resistance in the circuit), a maximum amount of lead is obtained: a quarter cycle, or 90° (remember that there are 360° in a complete cycle).

Because the voltage and current waveforms are no longer in phase, increased computational complexities exist when capacitors are used in ac circuits. This phase difference that exists between the current and the voltage is one of the main complications of ac circuitry.

Using a capacitive divider to measure capacitance

The test oscillator can be used to measure the value of a capacitor. The value of the capacitor that you made in Chapter 4 can be measured by setting up two capacitors as a capacitive voltage divider, as shown in

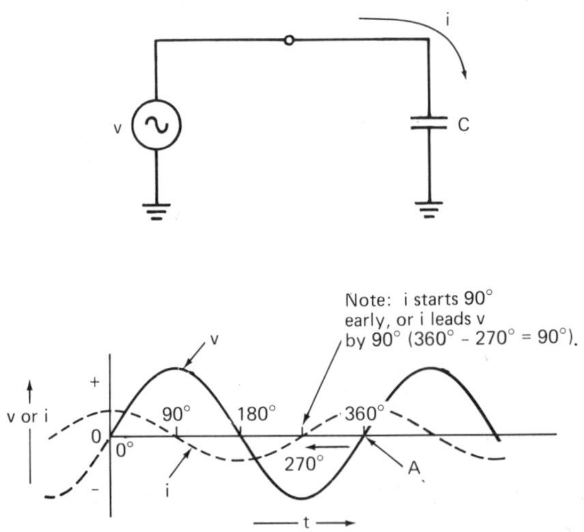

Figure 6.6 The phase relationship between current and voltage in a capacitor.

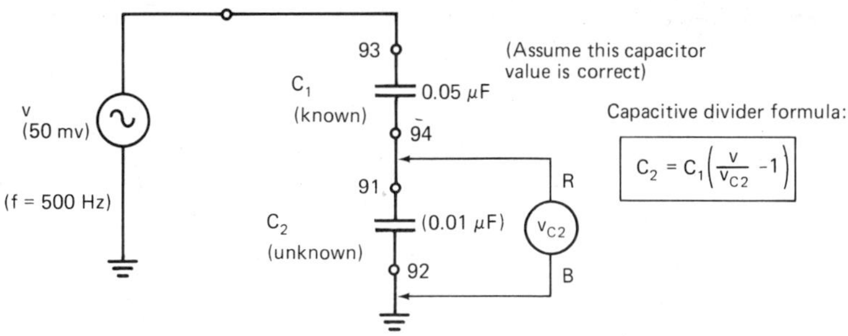

Example:
Assume you measured v_{C2} = 42 mv.
Then, using the capacitive divider formula

$$C_2 = 0.05\ \mu F\left(\frac{50\ mv}{42\ mv} - 1\right) = 0.05(1.19-1)\ \mu F$$

$$C_2 = 0.01\ \mu F$$

Figure 6.7 Measuring a capacitor using a capacitive divider.

Figure 6.7. The value of the upper capacitor C_1 has to be known in order to determine the value of the unknown capacitor C_2. The equation used is also shown on this figure.

To determine the value of C_2, v, the overall ac voltage that is being applied to the two capacitors in series has to be measured in addition to the value of the ac voltage that exists across the unknown capacitor C_2. When these two voltages are measured, the value of C_2 can be calculated.

To demonstrate and to also check out this measuring technique, the 0.01-μF capacitor of the lab kit will be used for C_2, the unknown capacitor, and the 0.05-μF capacitor will be used for C_1. The voltage across C_2 will be 42 mv (if 50 mv is supplied from the test oscillator). Putting these values of voltages in the appropriate places in the previous equation, as shown in Figure 6.7, the unknown capacitor C_2 can be computed to have a value of 0.01 μF, which is a good check on this measurement technique.

This technique will now be used to determine the value of the aluminum foil capacitor that you made in Chapter 4. Simply replace C_2 with your homemade capacitor, as shown in Figure 6.8. This time, a smaller voltage, about 28 mv, will be measured across C_2. Using the equation a second time, as shown in Figure 6.8, a value for C_1 of 0.04 μF is calculated (if you used plastic wrap for the dielectric).

For highest accuracy, keep the value of C_1 close to the expected value of C_2 so that approximately one-half of the applied voltage will appear across the unknown capacitor. This means that if either too small a voltage or too large a voltage is measured across C_2, make an appropriate change in C_1 so the voltage across C_2 is reasonably close to one-

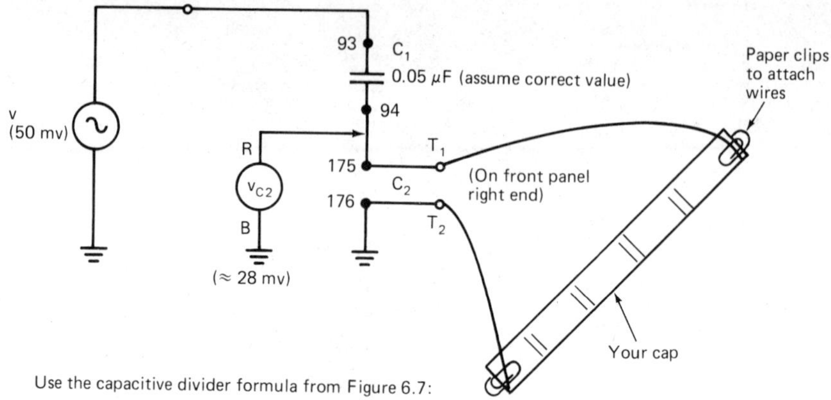

Use the capacitive divider formula from Figure 6.7:

$$\text{Your cap} = C_1\left(\frac{V}{V_{C2}} - 1\right)$$

$$= 0.05\left(\frac{50\text{ mv}}{28\text{ mv}} - 1\right)\mu\text{F}$$

$$\text{Your cap} = 0.04\ \mu\text{F}$$

Figure 6.8 Measuring your capacitor using a capacitive divider.

half of v. If the highest measurement accuracy is not needed, wider ranges in the voltage that is measured across C_2 can be allowed.

An *RC* low-pass filter

The test oscillator can also be used to experiment with a low-pass filter. In electronics, filters affect the output amplitude of signals in various ways that depend on the frequency of these signals. For example, filters are used to adjust the tone of the signals that come out of a high-fidelity audio system. The low frequencies can be boosted or attenuated, as can the middle range of frequencies and the high frequencies. In this way, the "coloring" of the sound coming out of the loudspeaker can be tailored to an individual's liking.

A low-pass filter lets low frequencies pass through with no attenuation, but it will attenuate high frequencies. The simplest low-pass filter is the *RC* circuit shown in Figure 6.9. A series 100-kΩ resistor and a shunt 0.05-μF capacitor to ground make up this filter.

Connect this low-pass filter to the output of the test oscillator, and then measure the resulting voltage that comes out of the filter (the ac voltage across the capacitor). Notice that although 50 mv of signal exists at the input to this filter, only 5 mv comes out. This is a 10 to 1 loss in voltage amplitude.

If a 0.001-μF capacitor is used, the output voltage would be higher because the capacitive reactance of this smaller-valued capacitor is

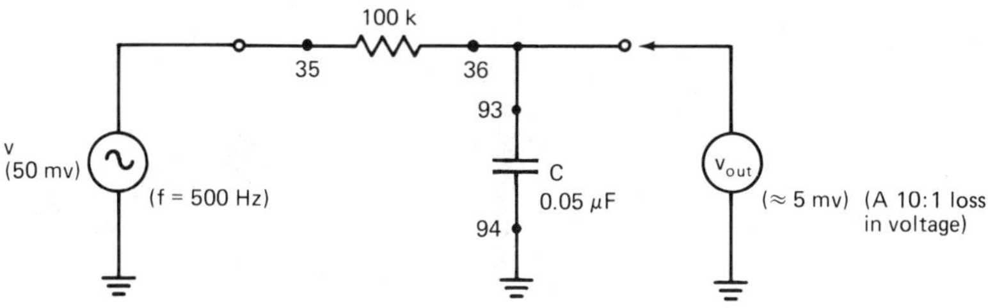

Note that if C = 0.001 μF is used, v_{out} = 46 mv (a very small loss in voltage).

But, if a dc input voltage is used, $V_{out} = V_{in}$ (unity gain).

Figure 6.9 An *RC* low-pass filter.

larger. Therefore, less signal is bypassed to ground. The output voltage would be about 46 mv, a relatively small loss from the 50-mv input signal level. Conversely, a larger-valued capacitor would cause a smaller output voltage to result.

If a dc voltage is used as the input to this low-pass filter, there is a "unity" transfer: the output voltage will equal the input voltage. Nothing is lost because low-pass filters pass dc (with this filter, the shunt capacitor to ground has no further effect once it charges up to the dc value of voltage).

Inductors

The discussion of inductors has been delayed until now so that some background could be developed about ac circuits. In modern electronics, circuit designers work hard to eliminate inductors because they can be large, heavy, and expensive. It is often more economical to use additional transistorized circuitry and a different circuit approach to eliminate the need for an inductor.

Many types of inductors are available; Figure 6.10 shows a representative sample. Some of these inductors are simply a coil of wire that uses a little plastic support structure to hold the wires in place. These are air-cored inductors. Other inductors, not shown in this figure, use a laminated core that is made up of thin sheets (and also even-thinner tapes) of special magnetic steel. These are the iron-cored and tape-cored inductors.

The electronic symbols used for inductors are shown in Figure 6.11. Part of the symbol indicates the type of core material that is used in the inductor. For example, Figure 6.11*a* shows the symbol for an air-cored inductor. In Figure 6.11*b*, the parallel dashed lines next to the coil signify a powdered-iron core. In Figure 6.11*c*, the solid parallel lines signify an iron core.

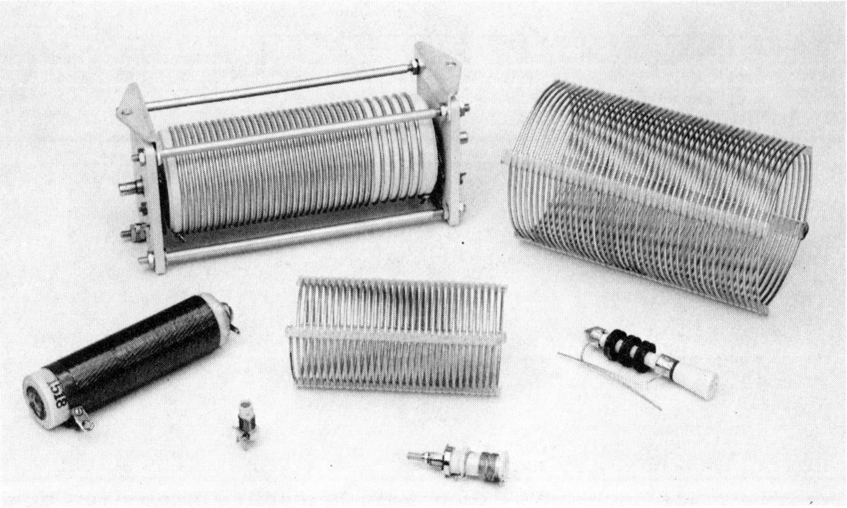

Figure 6.10 Inductors look like this. *(American Radio Relay League)*

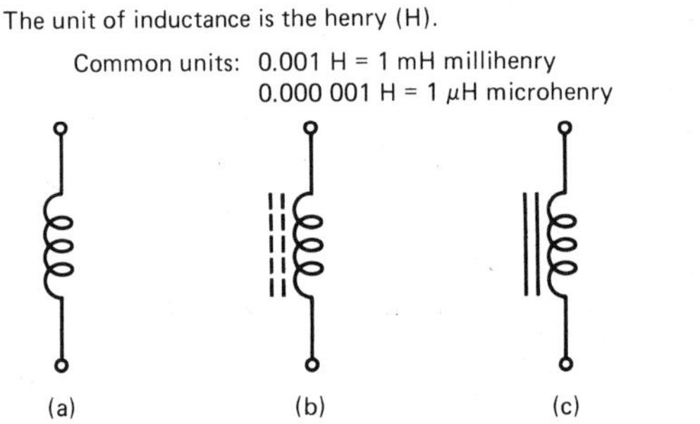

Figure 6.11 Inductor symbols. (*a*) An air-cored inductor, (*b*) an inductor with a powdered iron core, and (*c*) an inductor with a laminated iron core.

The unit of inductance is the henry and the smaller units of inductance are the millihenry (mH) and the microhenry (μH). What's considered a large value of inductance depends on the person who is using the inductor. As shown in Figure 6.12, audio enthusiasts and power supply designers consider large-valued inductors to be greater than 10 H. Digital system designers, on the other hand, consider that a millihenry is a large-valued inductor. Radio frequency engineers and circuit designers working with frequencies greater than 1 megahertz (MHz) consider 100 μH a large inductance value. At even higher frequencies, something like 100 MHz, 1 μH is considered a large-valued

Person	"Large" inductor values are greater than:
Audio enthusiasts and power-supply designers	10 henrys
Digital system designers	1 mH (0.001 H)
Radio frequency (RF) designers working at:	
1 MHz	100 μH (0.0001 H)
100 MHz	1 μH (0.000 001 H)

Figure 6.12 What is considered a large value of inductance?

inductor. The value of the inductors that are used depends on the frequencies of the ac voltages that exist in the application circuitry.

The core is important

Remember, from our discussions about capacitors, that if an air dielectric is replaced by a solid dielectric material, the value of the capacitance increases. Similarly, if a magnetic core is added to an air-cored inductor, a much larger value of inductance results, as shown in Figure 6.13. Adding a magnetic core to an inductor is therefore similar to using a higher dielectric constant material in a capacitor. It is a way to raise the value of inductance that can be obtained from a given coil of wire.

Iron cores increase the value of inductance but also introduce an additional loss of electrical energy because of the presence of this conductive core material. This is called the core loss. At low frequencies, laminating the iron into thin strips that are insulated from each other reduces this core loss. At higher frequencies, it is not practical to make

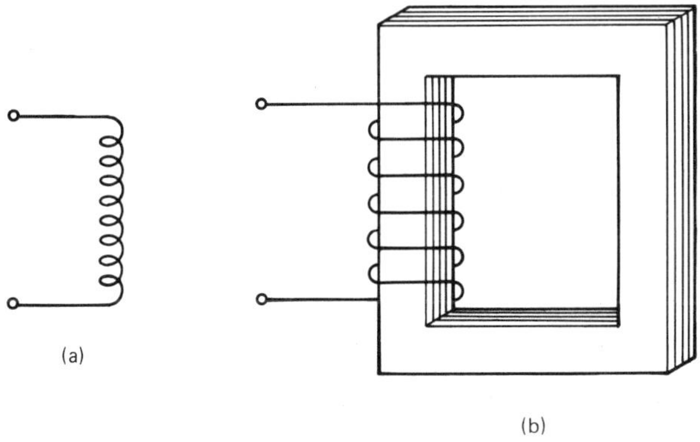

Figure 6.13 Adding a core raises the value of inductance. (*a*) An air core and (*b*) adding a core to increase the value of inductance.

the laminations thin enough to significantly reduce the core losses; therefore, both tape-wound cores and ferrite cores are used. A ferrite core is made from finely powdered iron that is combined with ceramic binders, compressed, and then sintered to produce a compact, solid mass of separated, insulated iron particles. Ferrite cores can be used at frequencies higher than the limit for tape wound cores.

Magnetic fields

Inductors are based on magnetic fields. When a permanent magnet attracts metal objects, this attractive force is said to result because of a "magnetic field" that is established by the magnet. Magnetic lines of "flux" are used to describe the strength of a magnetic field. These are diagrammed by the lines shown in Figure 6.14. The electrical property of inductance results from the effects of this magnetic flux. This magnetic flux exists with a single straight wire, but the flux is concentrated when the wire is formed into a coil. Coiling a wire therefore raises the value of inductance that is obtained.

As was seen earlier in this chapter, moving a permanent magnet over a coil of wire generates a voltage within that coil of wire. Moving the permanent magnet causes the magnetic flux to change, and it is this changing magnetic flux surrounding the wires that causes a voltage to be generated (called an "induced voltage").

Magnetic flux in an inductor depends on the amount of current that is flowing. The property of inductance exists because a change in the amount of current that is flowing causes a change in the magnetic flux,

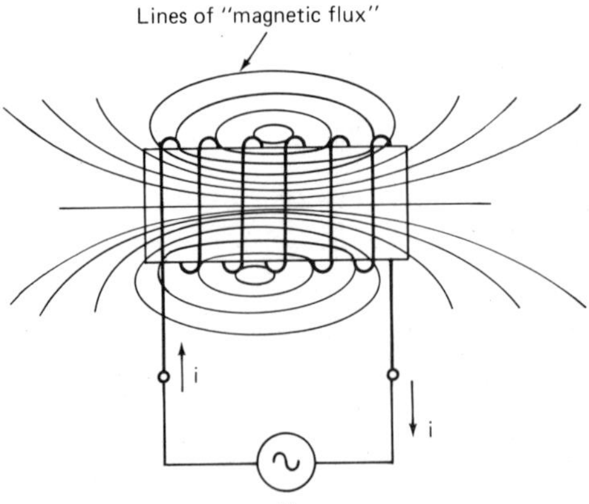

Figure 6.14 The magnetic flux that is established around a coil.

and this flux change induces or generates a voltage within the wire of the inductor. This induced voltage is of a magnitude that tends to resist any changes in the original current flow.

It takes energy to initially establish a magnetic field around an inductor, and this magnetic field, once established, resists any further changes. For example, if the current through an inductor tries to increase (as might result from applying an increasing voltage to the inductor), the increasing magnetic flux will induce an increasing voltage within the inductor that opposes the increase in the applied voltage and therefore limits just how rapidly the current flow can increase. Similarily, if the current flow in this charged inductor tried to decrease (as might result from applying a decreasing voltage to the inductor), the energy stored in the original magnetic field would generate a voltage within the inductor that also resists this change: the induced voltage would tend to keep the original current flowing. Therefore, current flow in an inductor tends to remain constant and it resists any changes. (This is similar to the voltage on a capacitor, which resists changes because the capacitor can charge or discharge to accept or to provide current in an attempt to hold the voltage constant.)

Once a current flow has been established in an inductor and then is abruptly interrupted, the terminal voltage of the inductor can "kick" (by magnetic induction) to a very high voltage, if necessary, in an attempt to keep the current flow constant. This "inductive kick" can cause surprisingly high voltages which can easily destroy transistors and ICs when circuit designers are not careful. Automobile engines make use of this inductive kick to increase the 12 V from the car battery to the high voltage that is necessary to "fire" the spark plugs. This will be considered in the section on transformers.

A demonstration circuit, Figure 6.15, shows how 1.5 V can be used to flash six LEDs in series, even though at least 9 V is needed to flash all these LEDs, because each LED requires approximately 1.5 V to conduct. When the Code key switch is open, the 1.5 V is not large enough to cause any current flow through the LEDs. Closing the Code key switch causes approximately 8.1 mA (the winding resistance is 185 Ω) to flow through the inductance of one winding of the input transformer. When the Code key switch is opened, this 8.1 mA continues to flow for a short time (to discharge the inductor) by flashing the LEDs. The voltage across the Code key switch can increase to a very large value, as a result of the inductive kick, if this large voltage was needed to maintain the current flow in the inductor. Special switches must be used to interrupt the current flow in inductive power circuits to prevent a continuous arc that would otherwise form across the switch and melt the metal of the switch contacts!

A "backswing diode" can be placed across the inductor, as shown in Figure 6.15*b*, to prevent this inductive kick. The 8.1 mA will now for-

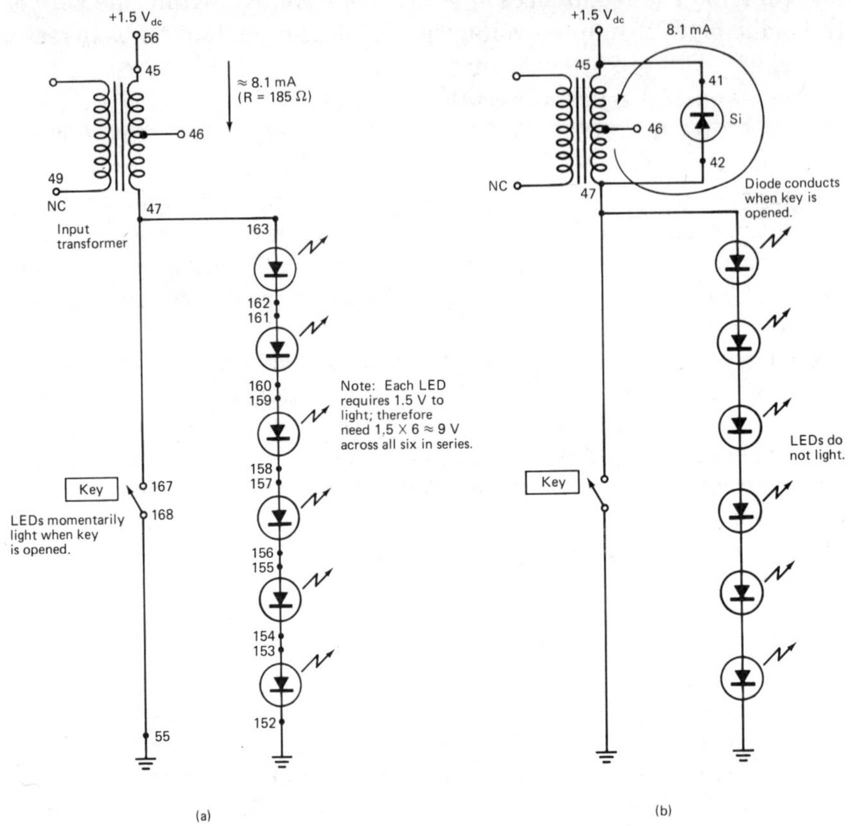

Figure 6.15 The inductive kick. (*a*) A demonstration circuit and (*b*) using a "backswing" diode to prevent the kick.

ward bias this diode when the Code key switch is opened, and this current will circulate around this local loop (the inductor and the diode) until the energy that was stored in the inductor has been dissipated. Add this diode and note that the LEDs no longer flash. Diodes are often used across the actuating coils of relays to prevent high-voltage breakdown of the transistor that supplies the energizing current to the relay coil.

This idea of inductance has also been used in commercial diesel-electric trains. When a train is going downhill, rather than using the brakes and dissipating all of this energy as heat, electric generators can be used as a brake. The energy of the train is used to generate electricity for use by other trains that might be coming up a hill.

Additional braking on electric trains is also provided by a coil of wire that is located in proximity with the surface of the metal track. This coil of wire induces current flow within the track. The magnetic field that is established by this induced current in the track tends to oppose the magnetic field that is creating that current flow. This is the principle of "magnetic braking"; it is transformer action.

Inductors in series and parallel

Figure 6.16 shows the total value of inductance that results when two inductors are connected in series or in parallel. Notice that inductors behave similarly to resistors. Two inductors in series provide a total inductance that is the sum of the inductances of both inductors, and two inductors in parallel provide a total inductance that is given by the product of the inductor values divided by the sum of the inductor values.

Inductive reactance

Inductors have a relatively low value of dc resistance. In the early chapters, the small values of dc resistance of the input transformer, the output transformer, and also the antenna coil of the lab kit were measured. When calculating how much current flows when a 1-H inductor is tied to the test oscillator, this dc resistance of the winding is not as important as the value of the inductive reactance. The dc resistance of most inductors is considerably smaller than the inductive reactance and can therefore usually be eliminated from consideration.

Inductive reactance increases directly with frequency. If a 1-H inductor is excited with 500 Hz, the inductive reactance is a little over 3 kΩ, as shown in Figure 6.17. If this inductor is connected to the test oscillator, 16 μa of current will flow.

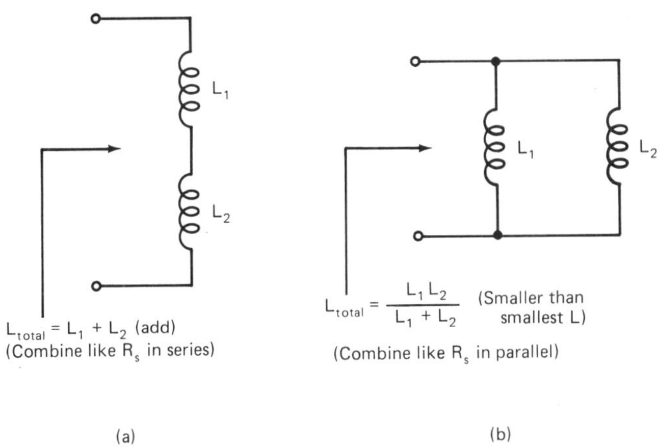

Figure 6.16 Inductors in series and parallel. (*a*) In series and (*b*) in parallel.

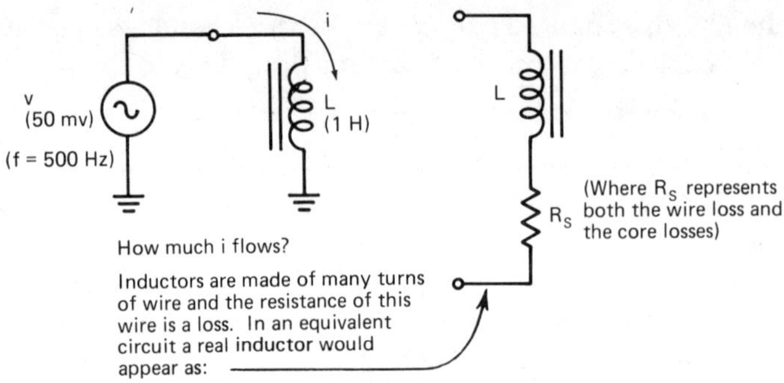

Because X_L is much greater than R_S, we can neglect the wire resistance. This inductor has an inductive reactance X_L given by

$$X_L = 2\pi f L$$

At the 500-Hz test oscillator frequency

$$X_L = 2(3.14)(500)(1\text{ H}) = 3.1\text{ k}\Omega.$$

Therefore, with v = 50 mv:

$$i = \frac{v}{X_L} = \frac{50\text{ mv}}{3.1\text{ k}\Omega} = \frac{5 \times 10^{-2}\text{ v}}{3.1 \times 10^{3}\ \Omega}$$

$$i = 1.6 \times 10^{-5}\text{ a} = 16\ \mu\text{a}$$

Figure 6.17 Inductors have inductive reactance.

Inductive reactance increases linearily with frequency. If the frequency is doubled, the inductive reactance is doubled, as shown in Figure 6.18. Inductive reactance therefore increases at high frequencies, and this is why people working with high frequencies typically don't need large-valued inductors.

The phase relationship between *v* and *i* in an inductor

When a sinusoidal voltage is connected across an inductor, a phase shift exists between the resulting ac current waveform and the ac excitation voltage waveform. Figure 6.19 shows that the current is lagging the voltage (rather than leading, as with a capacitor). When the voltage waveform is going through zero and increasing, it is 90° later in phase before the current waveform goes through zero and is also increasing. Therefore, the current lags the voltage by 90° in an ideal inductor. This is just the opposite of the phase relationship that exists between *v* and *i* in an ideal capacitor. If circuits are composed of resistors, inductors, and capacitors, current waveforms can exist that lead or lag by any amount from 0 to 360°.

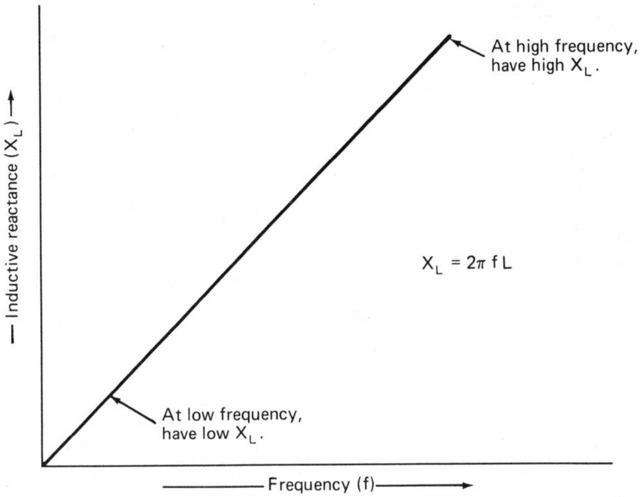

Figure 6.18 Inductive reactance varies with frequency.

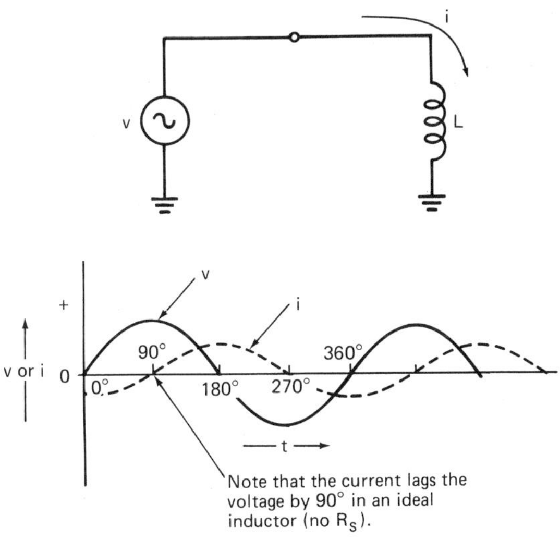

Figure 6.19 The phase relationship between voltage and current in an inductor.

Electrical resonance

When an inductor is connected in parallel with a capacitor, as shown in Figure 6.20, the resulting "tank circuit" has a very high impedance at the resonant frequency f_o (this resonant frequency is the particular frequency at which the inductive reactance is equal to the capacitive

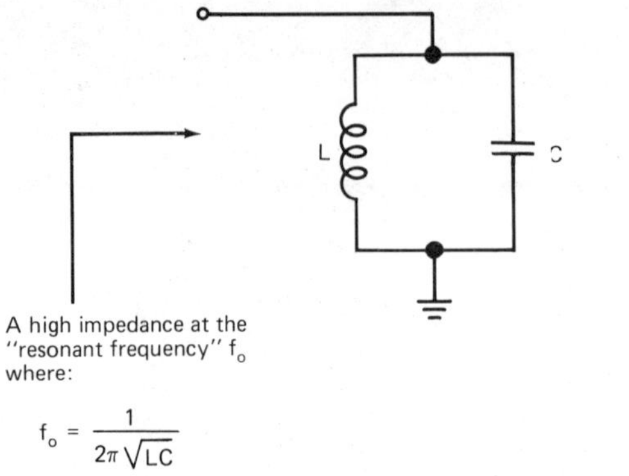

Figure 6.20 Electrical resonance: the tank circuit.

reactance). This high impedance results because none of the energy supplied by the excitation source is dissipated (with an ideal L and C, there is no R and so there are no power losses). Instead, this energy is exchanged between the magnetic field of the inductor and the electrostatic field of the capacitor. After an initial buildup of energy in this tank circuit, there is no more drain on the signal source, so a high impedance results.

Alternatively, if the L and C were connected in series as was mentioned earlier, a low impedance would result at the resonant frequency f_o. This results from the equation for the impedance of a series RLC circuit that was used to introduce this chapter.

Resonant frequencies also exist in mechanical systems. For example, when soldiers cross a bridge, they typically break cadence. They don't march across the bridge in step because the bridge might be resonant at this marching frequency and would then start swaying violently or could even come apart. Another example of a resonant mechanical system is a window in a home that rattles when a truck or a railroad train passes by. After the truck or train is gone, the window stops rattling because no other physical disturbance is providing the exact resonant frequency that is needed to excite that window. Tuning forks that are used to tune pianos are another example of a mechanical resonator. These forks provide accurate frequency references that are used by a piano tuner who then adjusts the tension on the strings that are provided so that each key will produce the correct frequency. In electronics, a "resonantor" can be made by connecting an inductor and a capacitor in parallel to form a tank circuit.

A tank circuit is used in the circuitry of the test oscillator and is composed of the 0.1-μF capacitor and the inductance of the complete winding of the output transformer (terminals 50 and 52), as shown in Figure 6.21. The frequency of this test oscillator is 500 Hz, and the capacitor that is

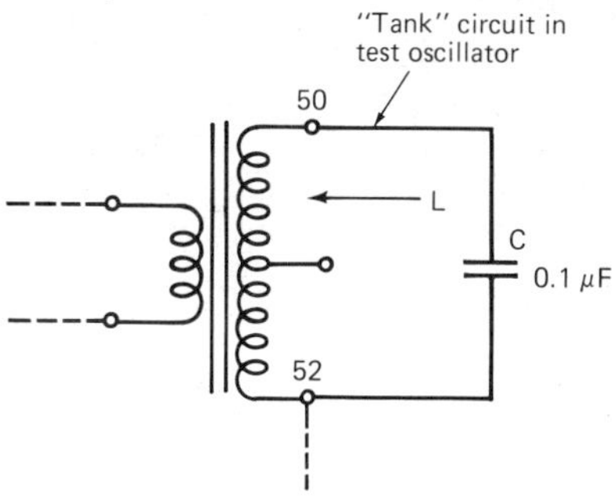

Note: $f_o = 500$ Hz

We can calculate L from

$$L = \frac{1}{(2\pi f_o)^2\, C} = \frac{1}{[(2)(3.14)(500)]^2\ 0.1 \times 10^{-6}}$$

$L = 1$ H

Figure 6.21 Calculating the inductance used in the test oscillator.

used in this tank circuit has a value of 0.1 μF. Therefore, as shown on the figure, the inductance of the transformer winding can be calculated to be 1 H. This 1-H inductor is *resonating* with the 0.1-μF capacitor to generate a frequency of 500-Hz for the test oscillator.

The antenna coil and the variable tuning capacitor can be used to build an oscillator circuit with a higher resonant frequency, as shown in Figure 6.22. An ordinary AM radio receiver can be used to measure this resonant frequency. This particular oscillator circuit makes use of transformer action between the antenna coil and the secondary winding of this coil that is available on terminals 69 and 70 to provide the reinforcing, positive feedback. Some of the output signal is returned to the input by use of this secondary winding to sustain the oscillation. Close the Code key switch to power this oscillator, and then you can receive the resulting radio frequency signal on an AM radio (the Code key switch must be held closed while you search for the signal by tuning the radio).

An easy way to detect this signal is to first tune the radio to an active station on which you can hear talk or music. Then slowly vary the Tuning control on the front of the lab kit (with the Code key switch held closed) until an audible beat frequency is heard coming out of the radio. This beat frequency will change in pitch as the signal from your radio station interferes with the received radio station. Leave the Tuning control at the point at which a high-pitched beat frequency is heard and then open the Code key switch; the interfering tone will go away. Every time the Code key switch is closed, this audible tone will be heard

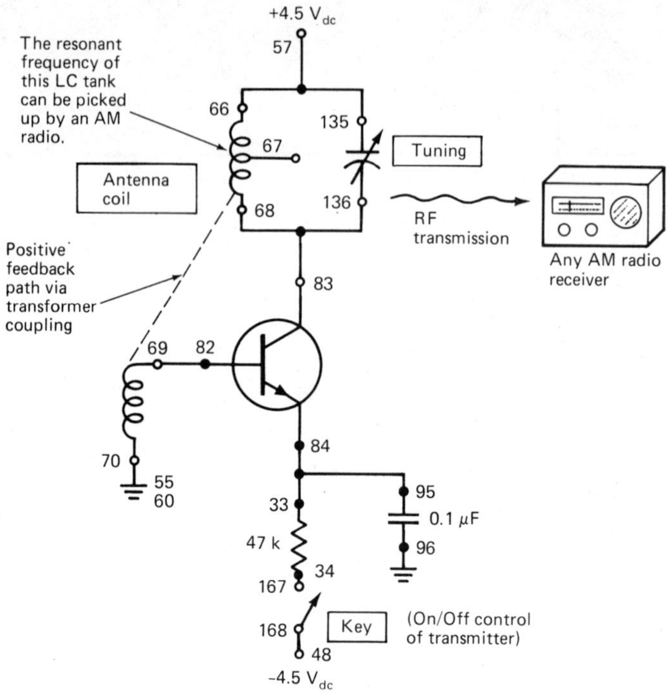

Figure 6.22 Building your own radio station.

coming out of the radio. By opening and closing the Code key switch rapidly, you can send Morse code that can be received by the radio. This high-frequency oscillator circuit is demonstrating a radio frequency resonance between a small-valued capacitor and a small-valued inductor.

An *LC* bandpass filter

LC parallel resonant networks can be used to make a bandpass filter, as shown in Figure 6.23. A bandpass filter results when a tank circuit is used in series with the collector of a transistor (the dc biasing circuitry that is needed for this transistor is not shown).

When a low-frequency input voltage is applied to this circuit, there will not be much of an output voltage provided from this bandpass filter. As the input frequency increases to values near the resonant frequency (f_o), there will be a corresponding increase in the output voltage. If the input frequency is increased further, a maximum output voltage will be provided at exactly the resonant frequency of the *LC* tank circuit. As the input frequency is then further increased beyond the resonant frequency, the output voltage response will again fall off. In this way, a particular frequency (the resonant frequency) will be

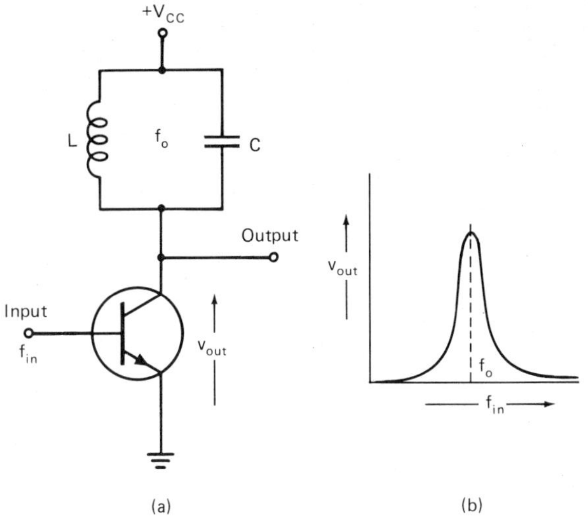

Figure 6.23 An *LC* bandpass filter. (*a*) Circuit and (*b*) response.

"selected" and all other frequencies will be rejected. This type of a resonant tank network is used in radio and television receivers to make sure that only one of the available stations (or channels) will be received, and it ensures that the other stations will not cause undesired interference. When a different station (or channel) is to be selected, the resonant frequency of the bandpass filter is changed so that this desired station is allowed to pass through.

An *LC* low-pass filter

In contrast to a bandpass filter, a low-pass filter will only pass low frequencies that are less than an upper cut-off frequency. Low-pass filters will also pass dc.

Figure 6.24 shows an *LC* low-pass filter. As the input frequency to this filter is increased, the inductive reactance of the series-connected inductor will also increase, and the capacitive reactance of the shunt capacitor will decrease. This causes the reduction in the output voltage at high frequencies that is shown in the figure. Inductors are still used in circuits and systems today, especially at high frequencies, to build filter circuits.

Transformers

The basic functioning within a transformer is very similar to the functioning of an inductor. If a second coil is brought in proximity with the

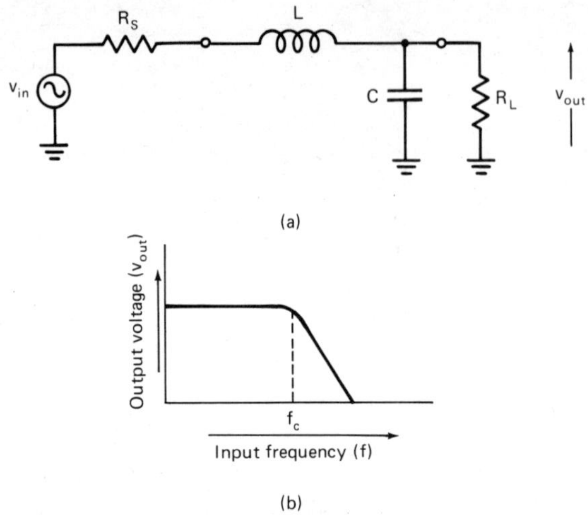

Figure 6.24 An *LC* low-pass filter.

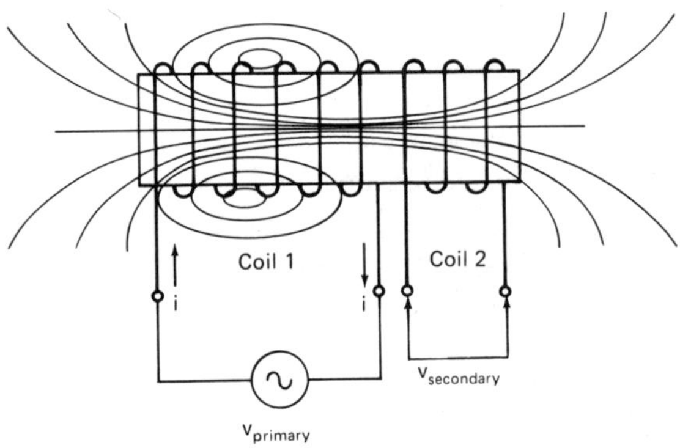

Figure 6.25 Inductively coupled coils: the transformer.

first coil, as shown in Figure 6.25, a transformer is created. This second coil also intercepts the lines of magnetic flux that are established by the first coil. This is called magnetic "coupling." The generation of a voltage in this "secondary" winding requires the changing magnetic field that is established by current changes in the "primary" winding. Therefore, transformers don't respond to dc (constant current) flow; an increase or a decrease in current flow is necessary (as is always happening with an ac current) to generate a voltage within the secondary winding of a transformer.

The basic construction of a transformer is shown in Figure 6.26. Two separated coils of wire are wrapped around the same laminated-iron

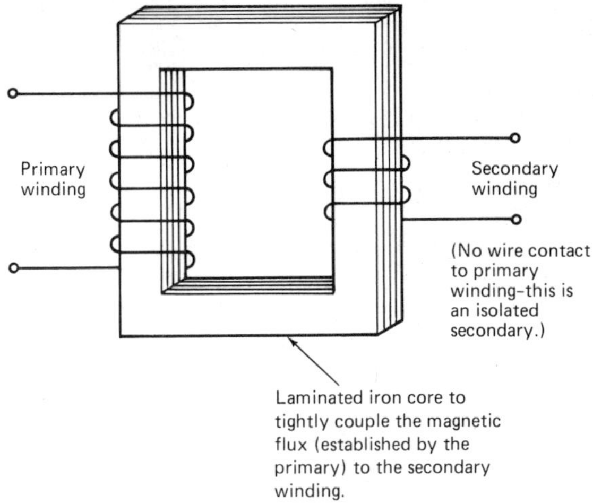

Figure 6.26 The basic construction of a transformer.

core. In this example there are more turns on the primary than on the secondary, so this is a step-down transformer; the secondary voltage will be smaller than the primary voltage.

One of the uses for a transformer is to provide electrical isolation. Neither terminal of the primary winding is in metallic or direct electrical contact with the secondary winding. There is therefore electrical isolation between the primary and the secondary circuits, even though these circuits are magnetically coupled (for ac signals).

Transformers look very much like inductors. The only way to determine whether a particular component is a transformer or an inductor is to count how many wire leads come out of the unit. Sometimes an inductor will have extra wires coming out that provide contacts to smaller sections of the winding so that the value of inductance can be varied. This is a tapped inductor and can be confusing, but tapped inductors can also be used as transformers.

A tapped inductor that is used as a transformer is called an "autotransformer." Autotransformers do not have electrical isolation because one contact of each winding is common. Autotransformers are typically lower in cost that a transformer because a separate coil doesn't have to be provided.

Some small-sized iron-cored transformers are shown in Figure 6.27. A close-up photograph of a miniature iron-cored transformer is shown in Figure 6.28. Notice that the laminations of the iron core can be easily seen. Iron-cored transformers are typically used in low-frequency applications, such as for power supplies (that operate at the 60-Hz power-line frequency).

Figure 6.27 Small-sized transformers look like this. *(Stancor Products, a Division of United Technology/Hamilton Standard Control Corporation)*

Figure 6.28 A close-up of a miniature transformer. *(Stancor Products, a Division of United Technology/Hamilton Standard Control Corporation)*

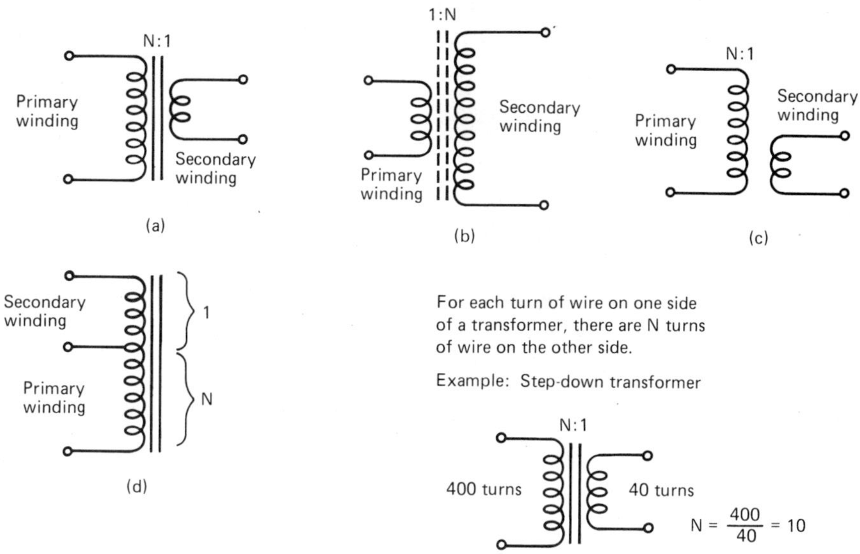

Figure 6.29 Transformer symbols. (*a*) A step-down transformer with an iron core, (*b*) a step-up transformer with a ferrite core, (*c*) a step-down transformer with an air core, and (*d*) an autotransformer (the secondary is not isolated).

The electronic symbols for transformers are shown in Figure 6.29.These are very similar to the symbols that are used for inductors. The type of core material is indicated in the same manner as with inductors. The turns ratio of a transformer is indicated as N (a number), and N:1 or 1:N is usually indicated next to the transformer symbol. For example, a step-down transformer with 400 turns on the primary and 40 turns on the secondary would have a turns ratio of 400/40, or 10:1. This means that there are 10 times more turns on the primary than on the secondary, so the voltage would be stepped down by 10:1. The turns ratio is an important parameter of a transformer.

Some electrical uses

There are electrical as well as electronic uses for transformers. Figure 6.30*a* shows a power-line pole transformer that is used to supply the 110-v electrical power for homes. The power companies typically distribute very high voltages because this allows small currents to flow in the distribution wires and still transfer relatively large amounts of electrical power. Power losses in wires depend on both the unavoidable resistance of the wires and the magnitude of the current flow. Raising the voltage reduces the current flow that is needed to distribute a given amount of electrical power, and this reduces the wire losses.

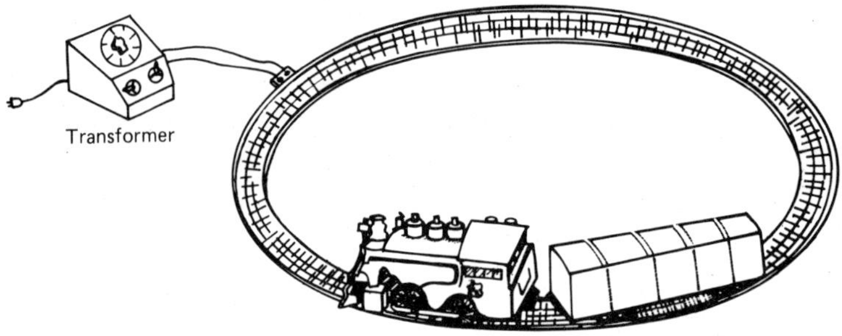

(b)

Figure 6.30 Some electrical uses for transformers. (*a*) A power-line transformer (*Pacific Gas and Electric Company*) and (*b*) a toy electric train transformer.

Another typical electrical use for a transformer is with a toy electric train, as shown in Figure 6.30*b*. Electric trains have a heavy small black box (the transformer) which usually has a knob on the top. This knob allows control of the secondary voltage by continuously changing the number of turns that a movable wiper picks up, much like in a potentiometer. Higher voltage is obtained with more turns of wire on the secondary, and this higher voltage makes the train go faster.

A doorbell transformer is used in your home to step the 110 v down to something like 12 to 28 v to operate a doorbell. Special transform-

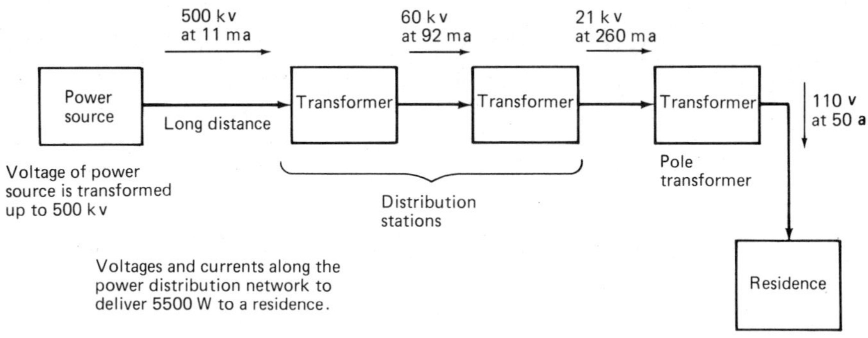

Figure 6.31 Using transformers in power distribution.

ers are also used to step 110 v down to approximately 45 v (which is safer to distribute in exposed areas that are close to the ground) to operate lights in home gardens.

An example of the voltages that are typically used in a power distribution system is shown in Figure 6.31. Voltages up to 500 kv are used on the long-distance lines. These high-voltage power distribution wires are supported by the tall towers that cross the country. Long insulators are used to attach these wires to the towers. Voltages of 500 kv are lethal; this is why there are the "HIGH VOLTAGE - DANGER" signs placed on each of these towers. These high voltages can cause electrical discharging into the surrounding air when there is high moisture content in the air. This "corona" discharging sounds something like a snake hissing, and it is certainly more lethal!

At the distribution stations this high voltage is first stepped down (using very large transformers) to 60 kv for further distribution. The next distribution stations then step this down to 21 kv for distribution within residential neighborhoods. A "pole transformer" (mounted near the top of the "telephone poles") is then used to finally step this voltage down to 220 and 110 v. Each of these lower voltage transformers will supply perhaps three or four houses. Notice that to provide 5500 W of power to a home requires 50 a at 110 v, but the high-voltage source of this electric power (a hydroelectric station, for example) needs to supply only 11 ma over the long distance, 500-kv, distribution wires to take care of a 5500-W load.

Some electronic uses

Some electronic uses for transformers are shown in Figure 6.32. Figure 6.32*a* shows a transformer that is used to change the magnitude of an ac voltage. This is similar to the use of the pole transformer in electric power distribution, only on a much smaller scale. In electronic sys-

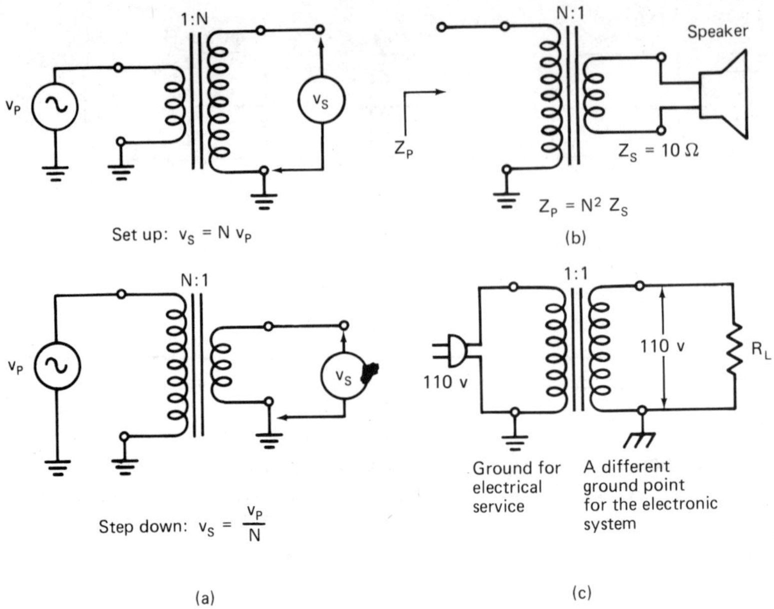

Figure 6.32 Some electronic uses for transformers. (*a*) Changing the magnitude of an ac voltage, (*b*) changing an impedance, and (*c*) isolating circuits.

tems, voltage levels are usually increased by using an amplifier rather than a transformer, but transformers can also be used (especially if the signal source can deliver some power).

In power distribution applications, a transformer is usually thought of as changing the magnitude of a voltage. In electronic applications, a transformer is often used to change the magnitude of an impedance or a resistance value (although the secondary voltage is still being changed, even in this application). Thinking in terms of the impedance change is just a different viewpoint. For example, in the lab kit the 10-Ω impedance of the loudspeaker is stepped up by use of the output transformer, as shown in Figure 6.32*b*. The higher resulting impedance level is a more optimum load for a transistorized amplifier.

Transformers are also used to isolate the electrical grounds that exist between different electronic systems, as shown in Figure 6.32*c*. Modern television sets and radios can create electrical shock hazards if a person comes in electrical contact with any of the metal screws that are embedded (to reduce the chance of contact) in the plastic cases. In the older television sets, power transformers were used and these transformers also provided electrical isolation from the power line. Today, these high-cost power transformers are usually not used, and therefore people should be careful not to contact any screws or internal metal parts of a TV set.

Checking the turns ratio

The turns ratio of a transformer can be determined by measuring the primary and secondary voltages. The ratio of these two voltages gives the turns ratio.

Figure 6.33 shows how to determine the turns ratio of the output transformer that is used in the test oscillator. While the test oscillator is operating, the secondary voltage is approximately 50 mv. The voltage across the total primary winding (between terminals 50 and 52) can be measured and is about 560 mv. The ratio of these two voltages (560 mv and 50 mv) therefore indicates a turns ratio of 11.2:1 for the output transformer.

The turns ratio of the input transformer, Figure 6.34, can be determined by exciting the primary winding with the test oscillator. A 10-μF capacitor is used to ac couple the primary winding to the oscillator so that a dc short circuit (the small resistance of the primary winding) is not placed across the output of the test oscillator. Note that 46 mv can be measured from terminal 45 to terminal 47. Move the ac voltmeter over to the secondary terminals 48 and 49, and read about 31 mv. The turns ratio of this transformer is therefore 46/31, or 1.5:1.

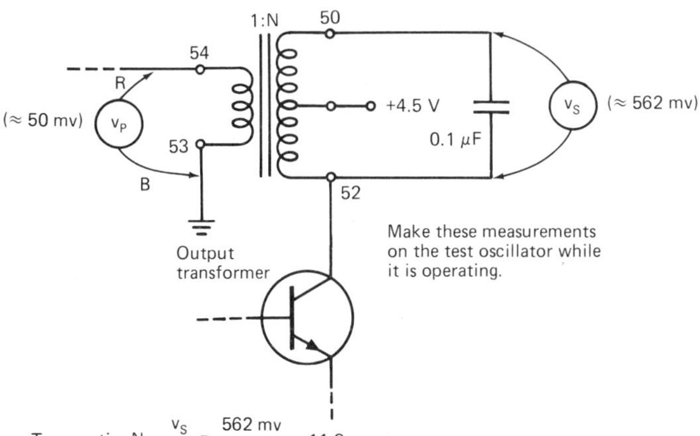

$$\text{Turns ratio, } N = \frac{V_S}{V_P} = \frac{562 \text{ mv}}{50 \text{ mv}} = 11.2$$

Figure 6.33 Checking the turns ratio of the output transformer.

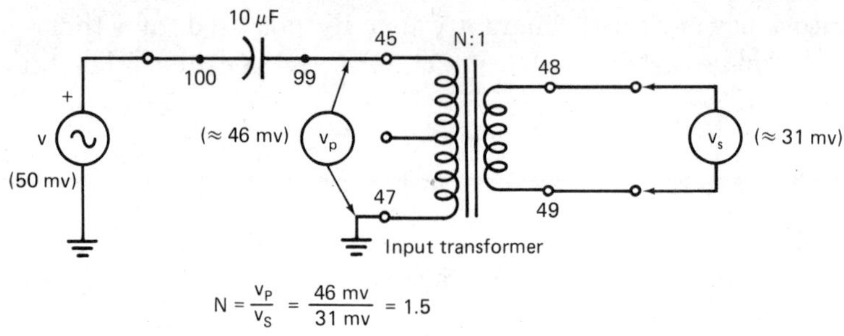

Figure 6.34 Checking the turns ratio of the input transformer.

The frequency response limits of transformer coupling

Remember that good capacitors don't pass dc. Also, transformers don't couple dc from the primary winding to the secondary winding. As shown in Figure 6.35, transformer coupling to a load resistance R_L can be represented by the equivalent circuit that is shown in Figure 6.36. The resistance R_L that is connected to the secondary is "transformed" and appears a factor of N^2 larger when "viewed" from the primary. In shunt (in parallel) with this transformed impedance is the primary inductance L_p of the transformer. At dc the inductor represents a short circuit, so no dc voltage is developed across the transformed load impedance. At higher frequencies, the inductive reactance of the primary, the "magnetizing inductance," increases and this then allows a voltage to develop across the transformed load. Obtaining a proper value of magnetizing inductance requires a particular number of turns of wire on the primary winding, and then the desired turns ratio determines the number of turns of wire that are needed on the secondary. So transformers with the same turns ratio are not equivalent because the magnetizing inductances will most likely be different. This creates replacement problems when transformers are used to transform an impedance because not only must the turns ratio be correct, but the frequency

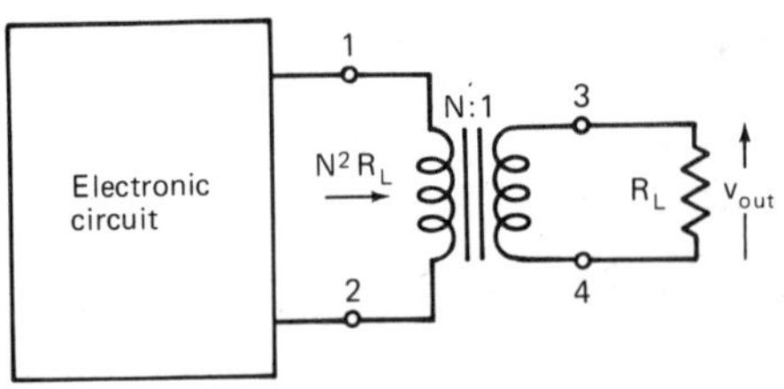

Figure 6.35 A transformer used to couple a load resistor to an electronic circuit.

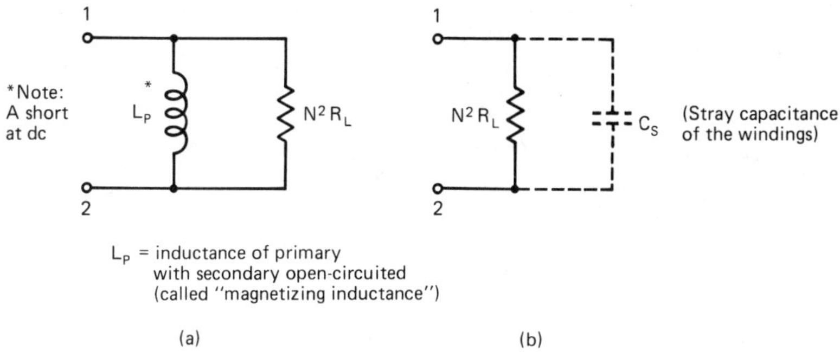

Figure 6.36 The equivalent circuit for transformer coupling. (*a*) At low frequencies and (*b*) at high frequencies.

response will change unless the magnetizing inductance and the stray winding capacitances (that we will consider shortly) are correct. We'll now take a closer look at the frequency capabilities of transformer coupling.

The operating frequency at which the inductive reactance of the primary is equal to $N^2 R_L$ (the transformed load impedance) is the low-frequency "corner frequency" of the transformer coupling to the load R_L. At this frequency, the voltage is 70 percent (−3 dB) of the maximum value that is obtained at higher operating frequencies. In hi-fi audio applications, for example, transformer coupling to the loudspeaker can restrict the bass response of the overall system.

The separate high-frequency equivalent circuit, shown to the right in Figure 6.36, indicates that the parasitic stray winding capacitance C_S of the transformer is in parallel with $N^2 R_L$. At high frequencies, C_S causes the response of the transformer coupling to fall off. At the operating frequency at which the capacitive reactance of C_S is equal to $N^2 R_L$, the high-frequency corner frequency is obtained. At this frequency, the voltage is again 70 percent (−3 dB) of the maximum value that was obtained at lower operating frequencies. Therefore, transformer coupling can also restrict the high-frequency response of an amplifier.

With transformer coupling, there are therefore problems at both low and high frequencies, as shown in Figure 6.37. The low-frequency response is limited because of the magnetizing inductance (L_P of Figure 6.36*a*) of the transformer, and the high-frequency response is limited because of the stray capacitance (C_S of Figure 6.36*b*) of the windings of the transformer. The useful frequency range of transformer coupling therefore lies between these lower and upper corner frequencies.

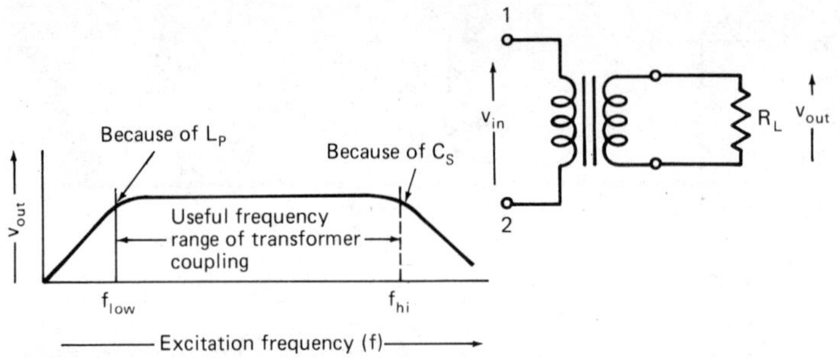

Figure 6.37 The frequency response of transformer coupling.

In older, vacuum-tube hi-fi systems, expensive transformers were used to obtain a broad bandwidth. The introduction of transistors (and, more recently, high-power audio ICs) has allowed the design of audio power amplifiers that directly (via a dc-blocking capacitor) couple to a loudspeaker. These circuits provide high-fidelity amplifiers at reduced cost because expensive output transformers are no longer used.

Now that all of this background material is in place, we'll take a look at the operational amplifier (op amp), the universal gain element that is the building block of linear systems.

Chapter

7

The Op Amp: The Versatile Linear Building Block

The op amp is the heart of most linear systems. This amplifier has two input terminals, a negative, or "inverting" input, and a positive, or "noninverting" input. A single output voltage responds only to the *difference* between these two input voltages. The output of an op amp can both source current to an external load or sink current (take current away) from the load. The symbol for an op amp is shown in Figure 7.1.

A very strange thing about an op amp is that there is no ground connection. Ground is not needed by the internal circuitry of an op amp but is usually introduced by the components (resistors and capacitors) that are used with an op amp. Other op amp terminals, not shown, are the pins that are sometimes made available to adjust the offset voltage (this will be discussed later) and the two pins that bring in the complementary power supply voltages (typically +15 and −15 V_{dc}). We will now take a closer look at an op amp so you can appreciate why it is so useful in many different application circuits.

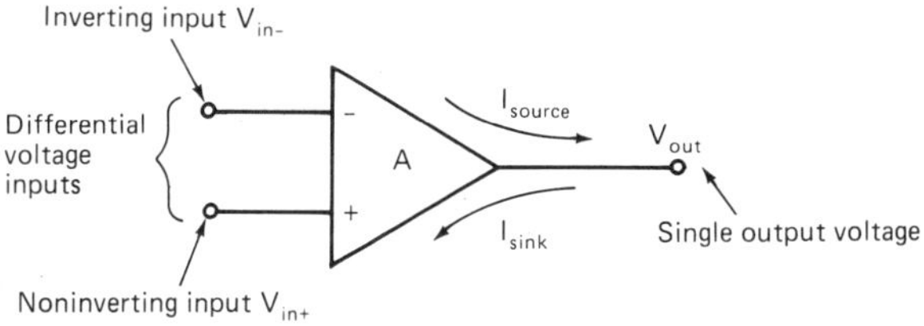

Figure 7.1 The symbol for an op amp.

The Voltage Gain of an Op Amp

Op amps have a very large voltage gain A_V (or simply A), as shown in Figure 7.2. Imagine grounding the inverting input, as shown in Figure 7.2*a* and then bringing a dc input voltage V_{in} into the noninverting input. We are interested in the output voltage V_{out} that results from this noninverting input voltage V_{in}. The input-output voltage transfer function shown in Figure 7.2*b* results. If V_{in} is exactly zero (the noninverting input is shorted to ground), an ideal op amp provides an output voltage of zero.

The output voltage of an op amp can swing either positive or negative, but it does not range quite as far as the actual magnitude of the power supply voltages. With plus and minus 15-V supplies, a typical op amp will provide a maximum output voltage swing of only plus and minus 13 V.

The horizontal scale, which represents V_{in} on the transfer characteristic shown in Figure 7.2*b*, is greatly magnified. Each major division is only 100 μV. If the input voltage is varied over the small range of ±130 μV, the output voltage will range over the complete span of

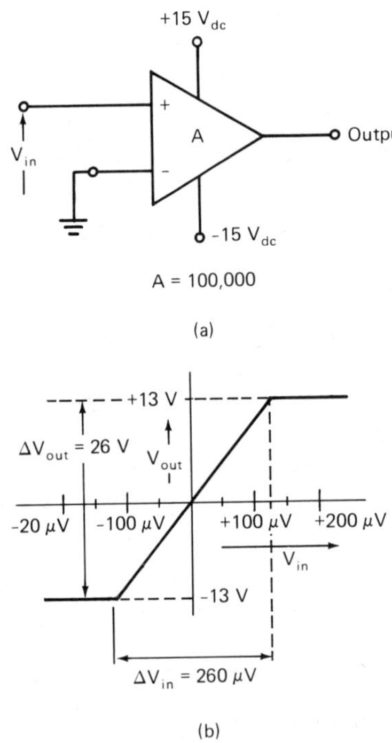

Figure 7.2 Op amps have large values of voltage gain. (*a*) An IC op amp and (*b*) the V_{in} and V_{out} characteristic.

−13 to +13 V. This corresponds to a voltage gain of 26 V/260 μV, or 100,000 (a fairly large voltage gain). Op amps are available with a voltage gain of over a million, and gains of 25,000 to 100,000 are very common with IC op amps.

From this transfer characteristic, it appears that V_{out} is usually at −13 or +13 V because if V_{in} is outside the range of ±130 μV, V_{out} is at ±13 V. After V_{out} goes up and stops at +13 V, the op amp is said to be "saturated." It is in positive saturation. Alternatively, if V_{out} had come down to −13 V and then "banged its head" (stops decreasing), the op amp would be in negative saturation. While the op amp is saturated, it quits behaving properly. But, if V_{out} is anywhere between this ±13-V interval, the differential input voltage (the difference between the two input voltages) is very small—typically no more than ±130 μV. In many cases, this 260 μV is small enough that it can be neglected. This gives rise to the important concept that *the two input voltages of an op amp are never very far apart.* Only very small-difference voltages need to exist between these two input voltages to provide any required output voltage.

An op amp mainly responds to the difference voltage that exists between the two inputs. If both inputs were at +5 V and then were both simultaneously taken down to −5 V (a 10-V change in the "common-mode input voltage"), the output voltage of an ideal op amp would not change. Voltages that are common to both inputs (common-mode input voltages) are rejected by an op amp. This ability of an op amp to amplify only the input difference voltage and not amplify the input common-mode voltage is indicated in a specification called "common-mode rejection ratio" (CMRR). This is the ratio of the magnitude of the differential voltage gain to the common-mode voltage gain. Large values of CMRR are desirable in op amp applications.

Feedback Is "Going Up an Attenuator"

To obtain an intuitive understanding of how amps are used, we will start by first considering the opposite of an amplifier, the voltage attenuator that was described in Chapter 4. A resistive attenuator that uses two 10-kΩ resistors is shown in Figure 7.3. With the 2-V input signal voltage shown, an output voltage of 1 V results (because the equal-valued resistors attenuate the input voltage by a factor of one-half). This takes care of the situation in electronic systems in which a large voltage exists, and something smaller is wanted. Notice that power supplies or amplifiers are not needed. When a voltage is too large and you want to make it smaller, this can be done simply with passive components—just a couple of resistors.

The current coming out of the tap of the resistors in the attenuator circuit is assumed to be zero; therefore there is no loading by any addi-

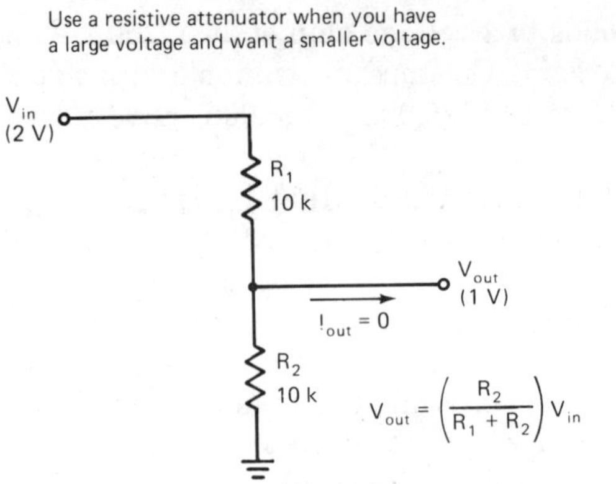

Figure 7.3 The resistive attenuator.

tional circuitry which may be connected to this tap point. No current is taken out of, or forced into, the resistive network at this tap point. If this current is zero (no loading), then the equation shown for V_{out} is correct. The amount of attenuation is only a function of the resistor values that are used. Therefore, the accuracy of the attenuation that is provided by this network depends only on the precision of these two resistors. If very precise resistors are used, a very precise attenuation results; there are no other errors with this resistive attenuator.

Often, the opposite problem exists in an electronic system in which instead of having too large a voltage, a small voltage exists that must be increased in magnitude. In this case, an amplifier is needed to amplify the small voltage. This means that some type of transistor circuitry and some power supplies are needed because it takes energy to provide amplification.

Figure 7.4 shows a diagram of an op amp application circuit that provides voltage gain. It is also seen to use a couple of resistors. These are the same resistor values that were used for the previous attenuator circuit. Now, instead of having 2 V and wanting only 1 V, we have 1 V and want to provide a 2-V output.

If the voltage at the tap point of these two resistors can somehow be forced to equal 1 V, the voltage across both resistors must be 2 V. So, you can look at this as "going up an attenuator." A voltage is going through the attenuator backwards: coming in on the low-voltage side and going out on the high-voltage side.

To accomplish this, the inverting input of the op amp is connected to the tap point of the two resistors, and the noninverting input is where the 1-V input signal is connected. To see how this works, start by assuming that the output of the op amp is initially at 0 V. If the output is at zero, the voltage at the tap point of these resistors (and therefore the invert-

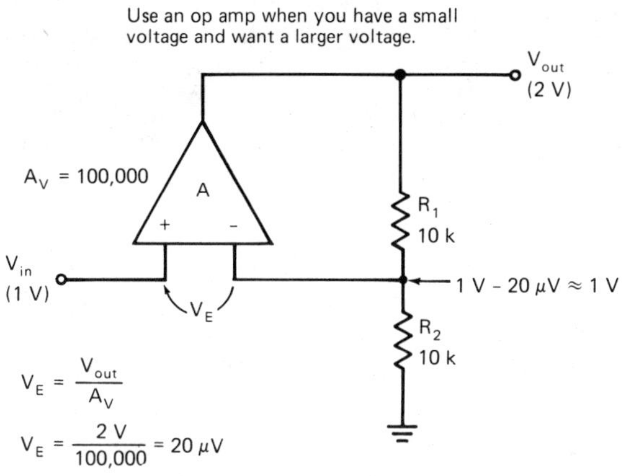

Figure 7.4 Feedback is going up an attenuator.

ing input of the op amp) is also at 0 V. This means that the 1-V input voltage appears as a 1-V differential input signal for the op amp ($V_{in+} = 1$ V and $V_{in-} = 0$ V). Now, the op amp has an open-loop voltage gain of 100,000; and, with a large differential input voltage (like this 1 V) the output voltage of the op amp is sent on its way to 100,000 V!

We have seen earlier that an op amp cannot provide output voltages as large as 100,000 V; so, instead, the output voltage would saturate at +13 V. But, the resistive attenuator is always sampling the output voltage; so, as the output voltage increases, the tap voltage (which is connected to the inverting input of the op amp) also increases because this tap voltage is exactly one-half of V_{out}.

This voltage at the tap point of the resistors cannot increase indefinitely because the op amp cannot allow this inverting input voltage to become larger than 1 V. For example, if the output voltage were to go too far positive, the inverting input would go to, say, 2 V, but the noninverting input voltage would still be at 1 V. The differential input voltage would then change to −1 V. This negative differential input voltage would make the output of the op amp go negative, contrary to our original assumption that the output voltage went too far positive: so this cannot happen.

What actually happens is that as the inverting input of the op amp gets very close to 1 V, notice that the *differential* input voltage to the op amp is disappearing. The action of the op amp is reducing its own differential input voltage. But, 1 V is still applied to the noninverting input, and as the inverting input voltage also approaches 1 V, a smaller and smaller differential input voltage results.

Assume that the output voltage will stop increasing when it reaches 2 V. Figure 7.4 shows that an error voltage V_E *would be needed to pro-*

vide this 2-V output voltage. Therefore, V_E will be the 2-V output voltage divided by the voltage gain of the op amp (100,000), or 2 V/100,000, which is 20 μV. This means that the inverting input voltage will actually stop increasing when it is just 20 μV short of 1 V. This is very close to 1 V because 1 V minus 20 μV is equal to 0.99998 V, and most people (and most voltmeters) would call this 1 V.

An equilibrium condition will therefore be established because this 20-μV differential input is now exactly supporting the 2-V output voltage. Because of the action of the resistive attenuator, the output voltage will actually be 40 μV smaller than 2 V (1.99996 V), but this amount of error is, again, difficult to detect. This small error is called the "gain error" of the op amp. If the op amp had a larger open-loop gain, say a gain of 1 million, this gain error would be 10 times smaller (2 μV), and then the gain error would be even less of a problem.

The resistors R_1 and R_2 are referred to as a "feedback network," and the resistor R_1 is connected from the output back to the inverting or negative input. This is called "negative feedback." Negative feedback is very useful in linear systems. In this example, the op amp will amplify voltages and the precision of the amplification does not depend on the characteristics of the op amp. The gain is determined by the values of the two resistors R_1 and R_2 that are used in the feedback network. The op amp was necessary to keep the gain error small, but any changes in the characteristics of the op amp—like unplugging one op amp and plugging in a different op amp—still provides an output voltage that will be very close to 2 V. Therefore, the performance specifications of op amps can have a wider tolerance than is allowed for the resistors that are used for the feedback network.

This op amp, with feedback, provides a noninverting voltage gain of very close to 2. The gain precision largely depends on the resistor values that are used in the feedback network R_1 and R_2. This is similar to an attenuator, but notice that providing gain is not quite as precise as providing attenuation because of the gain error.

This is one of the ways of using an operational amplifier. If the values of these resistors are changed, for example, if R_1 is made larger or if R_2 is made smaller, more voltage gain would be obtained. Whatever provides more attenuation in the two-resistor attenuator network that uses these same resistor values will provide more voltage gain in this fed-back op amp application. So when we "wrap these resistors around the op amp," we end up with a very predictable voltage gain. This is the magic of feedback and is the reason that system designers are willing to exchange the very large values of unpredictable and temperature sensitive "open-loop" voltage gain that an op amp provides for a smaller, very precise, and very stable "closed-loop" voltage gain.

The Basic Circuitry of an Op Amp

To appreciate how op amps work, let's take a look at some of the internal circuitry. As a learning exercise, you will then build a simplified demonstration op amp out of the components that are in the lab kit. The electrical characteristics of this op amp will be measured and then it will be used in a number of real-world application circuits.

The differential inputs of an op amp are provided by a differential amplifier, as shown in Figure 7.5. This uses two PNP transistors with their emitters tied together. The input voltages for the op amp are applied to the respective base leads of these two transistors. Notice that the input is a differential voltage and that the output will be the difference voltage V_2 that exists between the two collectors of these transistors. The same value resistor (1 k) is used for each collector load. Build this circuit to gain some understanding of how a differential amplifier works. This differential connection of two transistors is a basic building block in linear systems.

There is one large-valued resistor, 220 k, that is tied to both of the emitters in this differential amplifier. Measure the dc voltage across this resistor V_1; its value should be close to 4 V. This indicates that a total current (called "tail current") of 4 V/220 k, or 18 μA, is flowing. This tail current must either go into the emitter of the transistor on

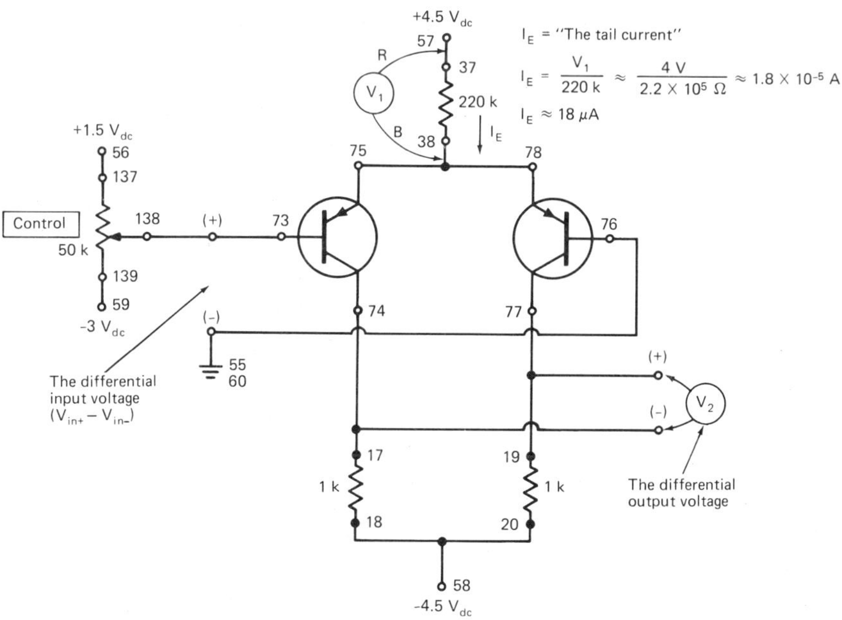

Figure 7.5 The input differential amplifier.

the left or into the emitter of the transistor on the right or be shared and go into both of these emitters simultaneously.

The steering of this tail current between the two transistors is controlled by the differentential input voltage that exists between the bases of these transistors. This is why an op amp only responds to the differential input voltage. For example, if the input voltage that is connected to the Control (the noninverting input) was at +0.5 V (and the other input is at ground), the transistor on the left will turn off and the transistor on the right then would carry the full 18 μA of current. This 18 μA essentially all comes out of the collector of this transistor and flows through the 1-k collector load resistor to produce an 18-mV voltage drop across this resistor.

If the positive input were connected, instead, to −0.5 V (and the other input was at ground), that would turn off the right-side transistor and turn on the left-side transistor. Now the 18 μA would flow through the left transistor.

If the dc voltmeter is connected between these collectors, various settings of the Control will provide either a zero differential output voltage (both transistors are conducting equal currents) or any voltage that lies between +18 and −18 mV. Build this circuit, use the Control to vary the noninverting input voltage, and observe this operation of a differential amplifier.

In an op amp, a differential output voltage is not provided. Therefore, a single output voltage (also called a "single-ended" output voltage) must be provided from this differential input amplifier stage. Also, load resistors generally are not used in the collectors of IC op amps. Instead, a particular connection of two transistors is used to replace both of these collector load resistors, as shown in Figure 7.6. Two NPN transistors, Q_3 and Q_4, are added to form a "current mirror": Q_3 is connected as a diode and Q_4 as a standard transistor.

To understand the functioning of this current mirror, first consider the balance condition (V_{in} = 0 V) where 9 μA is coming out of each of the collectors of Q_1 and Q_2. The 9 μA that flows to Q_3 automatically creates whatever base-emitter voltage drop is necessary for Q_3 to conduct 9 μA of collector current. But, this same base-emitter voltage is also applied to the base emitter of Q_4. This therefore turns on Q_4 to also carry 9 μA of collector current if these transistors are well matched in their characteristics, because they both have the same base-emitter voltage applied. Notice that when the input differential amplifier is balanced, the 9 μA coming out of the collector of Q_2 continues on into the collector of Q_4 and no output current will be provided from the current mirror: I_{out} is zero.

When you build this circuit, hook up the dc voltmeter across the 1-k resistor (as shown in Figure 7.6) as an indirect way to measure the output current of this circuit. The maximum output current provided

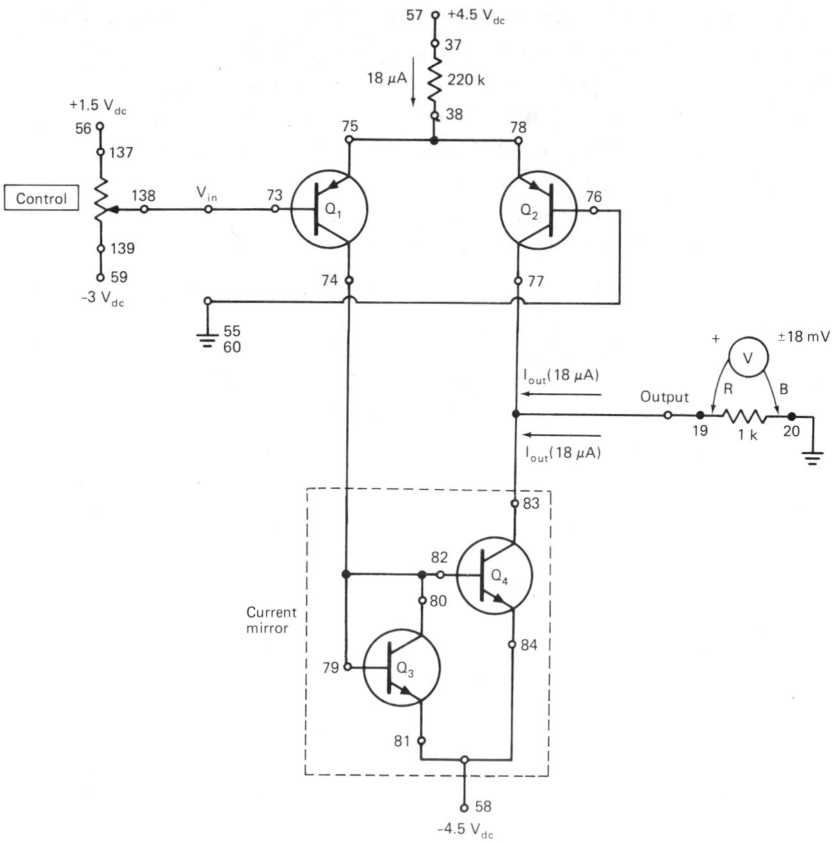

Figure 7.6 The current mirror replaces collector load resistors.

by the current mirror is 18 μA. Both the direction and the actual amount of this output current depend on the differential input voltage.

Adjust the control for an output current of zero. Now you can move the red lead of the dc voltmeter over to measure V_{in} at the base of Q_1 (terminal 73). This is the differential input "offset voltage" V_{OS} that achieved this balanced condition. An ideal op amp has an offset voltage of zero, but real op amps have an offset voltage that can be as large as 10 to 20 mV for low-cost op amps (although more expensive op amps are available with offset voltage specs of only a few microvolts).

Notice that this input differential amplifier and the current mirror are converting the differential input voltage into an output current that can either be supplied to, or absorbed from, the next stage of the internal op amp circuitry. Therefore a bidirectional output current is provided by this input stage.

This basic type of input stage is used in many IC op amps. Understanding the operation of this circuit therefore provides a very good idea of the internal operation of an op amp. The actual schematic of an IC op amp includes many more transistors and it is usually difficult to see what's really going on, but this basic mechanism takes place in most op amp input stages.

We can add to this input circuitry and provide a very good idea of what a complete op amp looks like in a simplified basic form, as shown in Figure 7.7. This same basic input stage is used, but a change has been made. Rather than using a resistor in the common emitter lead of the differential amplifier, a current source (labeled 2 I_E on the figure) has been added. A current source provides an op amp with better performance because the common-mode dc input voltage to the op amp can be changed without affecting the magnitude of the tail current. Remember that the current provided by a current source is essentially independent of the voltage that exists across the current source. If a resistor were used in place of this current source and both dc input

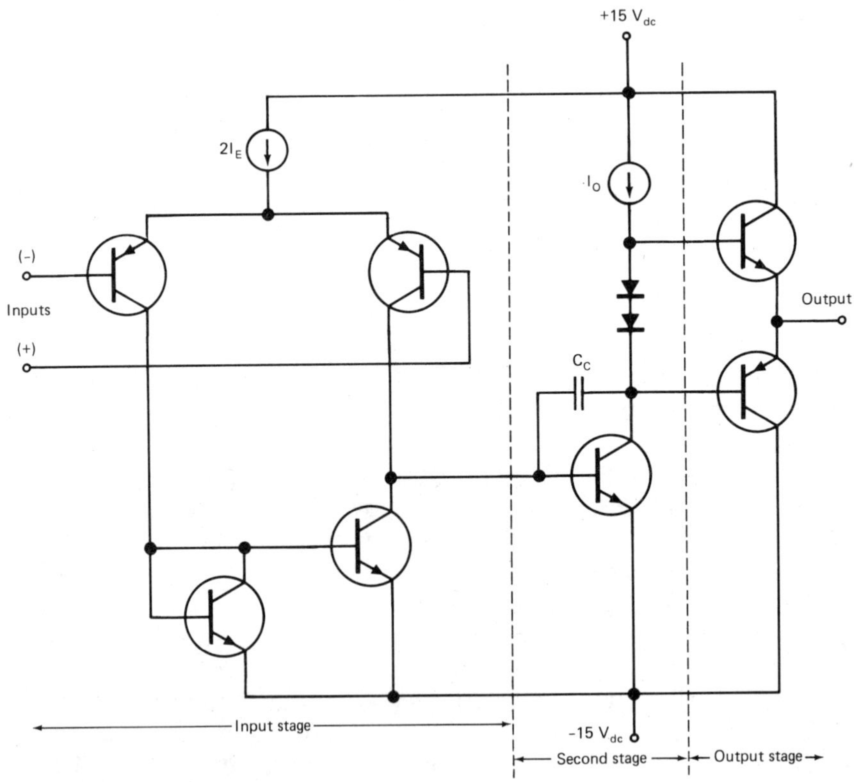

Figure 7.7 A basic op amp.

voltages were increased, say up to a common-mode level of 10 V, the voltage drop across this emitter resistor would be reduced and therefore the tail current would also be greatly reduced. This current source provides an independent tail current that essentially will not change as a result of common-mode input voltage changes.

A second stage common-emitter amplifier is now added to the input stage. This common-emitter amplifier has a current source, labeled I_0, in its collector. The high output impedance of a current source provides much higher voltage gain than can be achieved with a collector load resistor.

There are a few additional components added to this second stage; there are two diodes between the current source and the collector lead. These diodes are used to bias the output stage, the upper NPN transistor, and the lower PNP transistor (this output stage will be discussed next). In addition, there is a "frequency compensation" capacitor C_c that goes from the base to the collector (from the input to the output) of this second stage. This capacitor is typically provided on the IC op amp chip and is used to prevent undesired oscillations in op amp application circuits. Typically a large-enough capacitor is provided by the IC manufacturer so that any amount of feedback can be used without having problems with undesired oscillations. Therefore, only some unusual application circuits will give oscillation problems with modern IC op amps.

Some op amps are available in which this "comp cap" has to be supplied by the customer, an "externally compensated" op amp. This has advantages because the customer can then use exactly the value of capacitor that is needed, and this will extend the bandwidth, especially in high-gain application circuits. This adds to the overall cost of each application circuit because this additional capacitor has to be provided. Modern "internally compensated" op amps (which have a large-enough value for the comp cap on the chip so that any amount of feedback can be used) provide performance which is quite adequate for most applications, and they are more popular than externally compensated op amps.

The output stage is a complementary emitter-follower amplifier. It is called "complementary" because both transistor types are used, an NPN on the upper side and a PNP on the lower side. The upper NPN transistor is the device that will provide the current that goes out the output terminal of the op amp and into a load resistor (the current that is "sourced" by the op amp). The PNP absorbs, or "sinks," current so that the output of this op amp can both source and sink current to an external load.

The two diodes that are in the collector circuit of the second stage are used to keep both of these output emitter-follower transistors slightly conducting under quiescent conditions. It is desirable to have a small current flowing through both of the emitter follower transistors of the output stage. Then, as current must be supplied to an external load, either the upper or lower transistor can quickly satisfy these demands for extra current because both of these output transistors are slightly conducting and can therefore more rapidly carry the additional current that is needed by the load.

Most IC op amps can be broken down into the same three basic internal stages that are shown for this very simple circuit. If there were enough transistors in the lab kit, this op amp could be built, and it would provide good performance. Since there are not enough transistors in the lab kit, a simpler op amp will be built. It will not perform as well, but it will still demonstrate the basic operation of an op amp and will provide a "feeling" for the operations that are taking place within the internal circuitry of op amps.

Making an Op Amp with Kit Parts

The circuit diagram for this demonstration op amp is shown in Figure 7.8. The two PNP transistors in the lab kit are used for the input differential amplifier. One of the NPN transistors is providing the second stage and the other NPN transistor is providing the output emitter follower. The output of the op amp is provided at the emitter of this emitter follower. This is not a complementary output emitter follower; it is only the upper half. The 10-k resistor in the emitter of Q_4 allows this op amp circuit to sink current, but it will not sink as much current as it would if an additional PNP for the lower side of this stage were available.

A couple of other simplifications have also been made. A current source is not used in the emitter of this input differential amplifier. It has been simplified to use a 220-kΩ resistor. The collector load for the second stage has also been simplified to use a 100-kΩ resistor. Also, since there are not enough transistors to make a current mirror for the input differential amplifier, one 47-kΩ collector load resistor is used.

The compensation capacitor has been made ridiculously large, 0.1 μF. This is much larger than is necessary (IC op amps use from 3 to 30 pF), but it provides a very large amount of "overcompensation" and will guarantee that there will be no undesired oscillation problems. This is therefore a very slow-responding op amp that won't give any high-frequency troubles.

This demonstration op amp can be built out of the components that are available in the lab kit. It operates from ±4.5-V power supplies, and it is a real, functioning op amp. With this basic op amp, many application circuits can be experimented with and the voltmeter can be used to measure the voltages within the op amp as well as all of the external voltages of the application circuits. This will assist your understand-

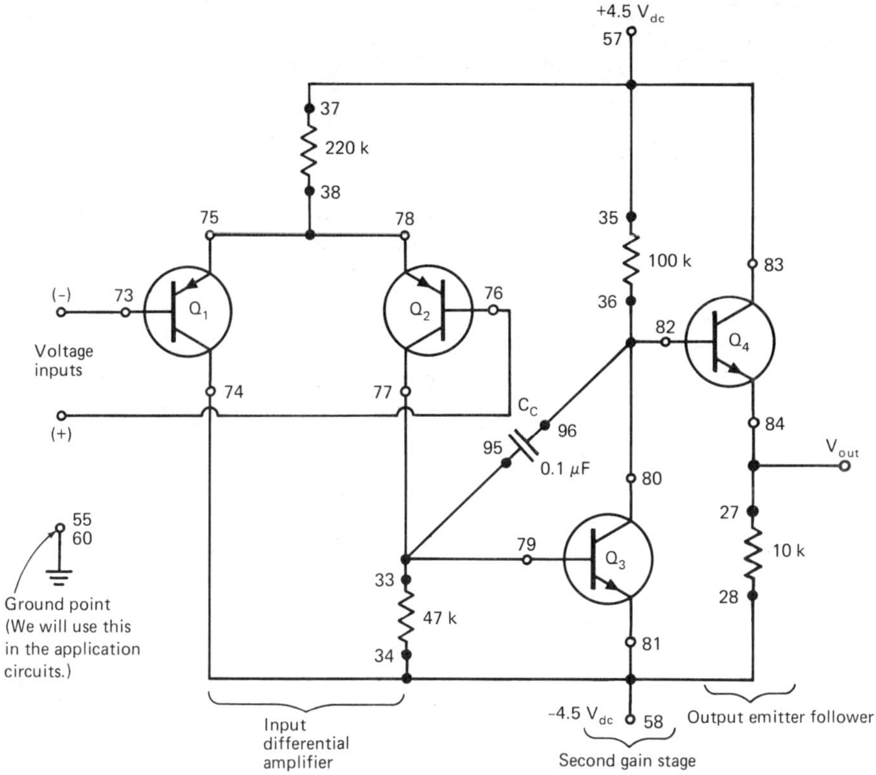

Figure 7.8 Making a demonstration op amp with components from the lab kit.

ing of how they work. It is a very interesting way to get some very solid feelings about op amp applications and also about the internal circuitry of op amps. This op amp will now be used to demonstrate a number of the popular application circuits, so don't take it apart.

Testing Your Op Amp

Before we start looking at some of the applications for op amps, let's spend a little time making some measurements on the demonstration op amp to get a basic idea about the specifications that are provided on op amp data sheets.

Measuring the offset voltage

The offset voltage of an op amp is defined as the differential input voltage that is needed to make the output voltage of the op amp go to 0 V. The offset voltage can be either a positive or a negative voltage. In the

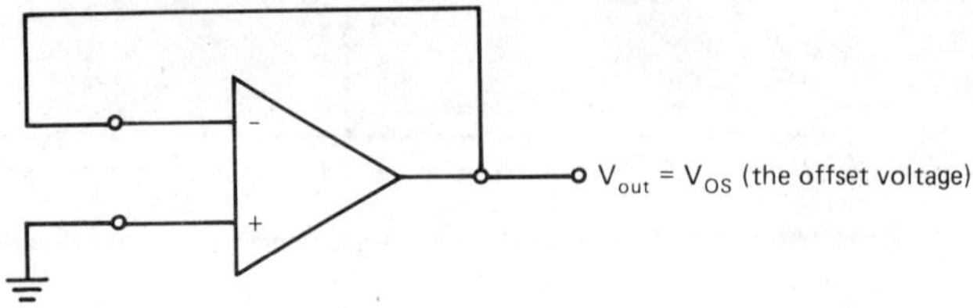

Figure 7.9 Measuring the offset voltage of the demonstration op amp.

specifications for op amps, there is no concern about the algebraic sign of the offset voltage. A good op amp design will have a random distribution of the offset voltage of the manufactured products about the ideal value of 0 V.

A guaranteed low offset voltage is desirable for an op amp. It is then not necessary to externally adjust the offset voltage to zero. This offset voltage zeroing adds to the costs because it requires a potentiometer and someone must adjust this potentiometer during the electrical testing of each circuit.

The offset voltage can be measured using the circuit shown in Figure 7.9. The noninverting input is simply connected to ground and a wire is connected from the output of the op amp to the inverting input (this provides 100 percent feedback). This introduces a small error because the output voltage is not exactly zero (as it should be in the definition of offset voltage). The output now sits at the offset voltage.

Measuring the input currents

Another specification of an op amp is the input current that is required. This input current must be supplied by the circuitry that the op amp is connected to. In application circuits, it is very desirable to have very low input currents. This is like the digital multimeter. This meter has an input resistance of 10 MΩ and therefore does not draw very much current and therefore does not significantly change the voltage at the nodes that are measured. Op amps should also have very low input currents so that they do not load the signal sources with which they work.

The input currents of this demonstration op amp are fairly small and therefore the indirect method of measuring current will be used, as shown in Figure 7.10. Notice that the largest resistor in the kit, 470 kΩ, is used. As shown in Figure 10*a*, the output of the op amp is shorted to the inverting input and the 470-kΩ resistor is placed in series with the noninverting input lead before this lead connects to ground. The dc voltage drop across this resistor is used to calculate the input current I_{in+} that is flowing at the noninverting input of the op amp. This reading should be about 23 mV, which corresponds to an input current of 49 nA.

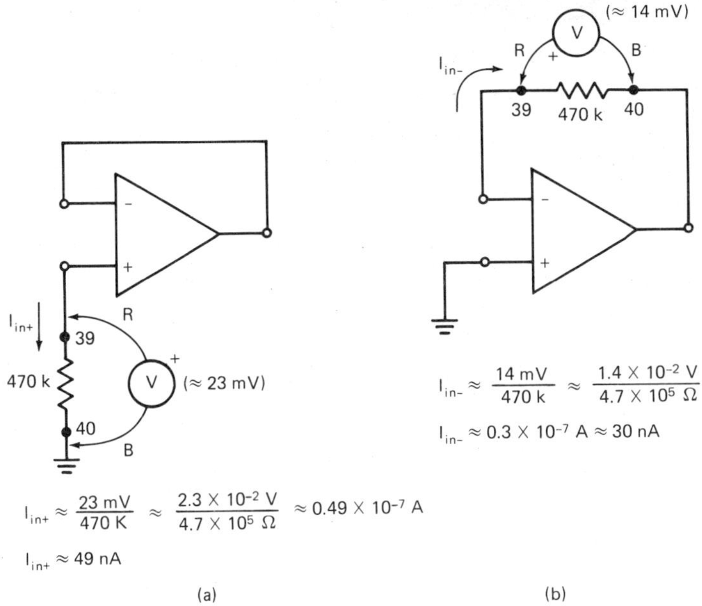

Figure 7.10 Measuring input currents of the demonstration op amp. (*a*) Circuit connection for I_{in+} and (*b*) circuit connection for I_{in-}.

This input current is actually coming out of the input lead because the input differential amplifier uses PNP transistors. In actual IC op amps, both PNP and NPN transistors are used directly at the input; therefore input currents can either come out of, or go into, an op amp.

Figure 7.10*b* shows how to measure the input current at the inverting input. The 470-kΩ resistor now replaces the direct wire that was used between the inverting input and the output of the op amp. This inverting input current is about 30 nA.

The difference between these two input currents is called the "input offset current." In this example, the input offset current is 49 − 30 nA, or 19 nA. If everything in this op amp circuit were perfectly symmetrical, the noninverting input current would then equal the inverting input current and the offset current would be zero.

Measuring the large-signal open-loop voltage gain

The open-loop voltage gain of an op amp is another very important specification. "Open loop" means there is no feedback network in place; this is the gain of the op amp by itself. With this demonstration op amp, the low gain (that has resulted from the simplified circuit that has been used) is relatively easy to measure, as shown in Figure 7.11. A small

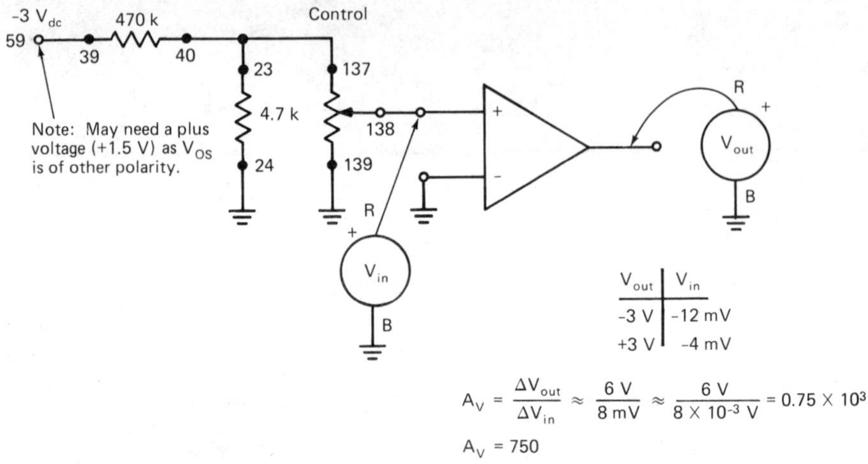

Figure 7.11 Measuring the large-signal open-loop voltage gain.

input signal is provided for the op amp by making use of the 470-kΩ and the 4.7-kΩ attenuators shown at the extreme left side of this diagram. The 4.7-kΩ resistor is then shunted with the potentiometer (Control) to get a variable amount of voltage to be used as the dc input voltage for the op amp. (If an IC op amp were to be measured, this technique wouldn't work and a more complex testing circuit would have to be used.)

A small voltage is desired as the input to this attenuator. Either the −3-V or the +1.5-V tap on the battery pack may have to be used, depending on the sign of the offset voltage of your op amp. Change to +1.5 V if −3 V doesn't work with your op amp.

Vary the Control to first provide a −3-V output from the op amp. Use the dc voltmeter at the output of the op amp to measure this output voltage. Now, shift the dc voltmeter to the noninverting input and measure and record the input voltage V_{in} that was necessary to provide the −3-V output. An input voltage of approximately −12 mV should be found. Then, change the Control setting to provide a +3-V output from the op amp. Again, move the meter lead over to the noninverting input and measure an input voltage of approximately −4 mV. The "large signal" voltage gain is the change in output voltage divided by the change in input voltage that was needed to obtain it. Moving from −3 to +3 V is a 6-V change in V_{out}, and the difference between the input voltages of −12 and −4 mV is 8 mV. So the voltage gain is 6 V divided by 8 mV, or 750. This is a considerably smaller voltage gain than the 100,000 that we have been previously using for a typical

IC op amp. This is another one of the performance characteristics that suffers because of the very simple circuitry that is used for this demonstration op amp.

Building Some Application Circuits with Your Op Amp

Now that we've looked at some of the ways of making basic measurements to determine the parameters of this demonstration op amp, we'll use it to become familiar with many of the popular application circuits. Whenever you first consider electronic circuits, you will find that there is generally a whole new language involved and one of the bigger problems is simply knowing what all of the words mean so you can understand what others are talking about. After you are introduced to this new vocabulary, you will find that the basic physical principles involved are used time and time again in different ways to achieve a wide range of useful electronic system functions.

A voltage follower

A very simple application for an op amp is the voltage follower shown in Figure 7.12. This is similar to the emitter follower that was discussed earlier, but this is a more precise buffer amplifier. It has a much higher input resistance because the input current of an op amp is typically smaller than the base biasing current of a transistor.

Notice that the output is directly connected to the inverting input. This is 100 percent feedback. A unity voltage gain amplifier results because whatever voltage is applied to the noninverting input will appear at the output of the op amp. The dc output voltage will differ from the dc input voltage only by the few millivolts of offset voltage of the op amp (plus the effects of the op amp gain error: the additional

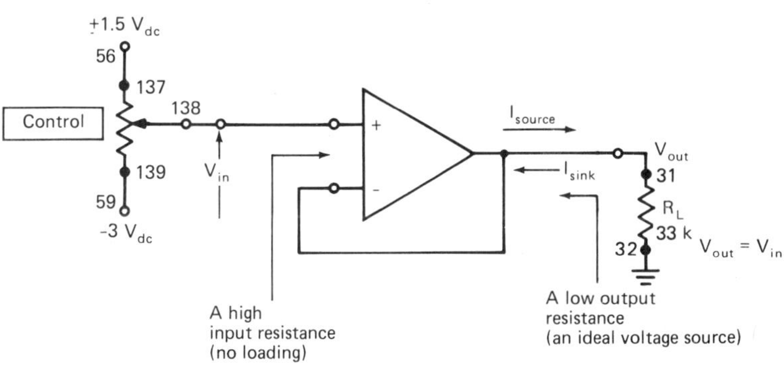

Figure 7.12 The voltage follower.

few microvolts of input voltage that are necessary to provide the output voltage). Therefore the match between the dc input voltage and the dc output voltage is much better with this voltage follower than can be obtained with an emitter follower. For an NPN emitter follower, the dc output voltage is lower than the dc input voltage by 600 or 700 mV; but, in an IC op amp voltage follower, the dc output voltage is essentially identical to the dc input voltage.

Voltage followers are very good isolation amplifiers; they have a high input resistance and a low output resistance. An op amp voltage follower can be used to eliminate circuit loading because the input current that is needed by the op amp is very small. An op amp can also provide whatever current is demanded by a load resistor R_L that is connected to the output. For most IC op amps, R_L should be no smaller than 2 kΩ.

An amplifier with noninverting gain

An application that uses an op amp to provide noninverting voltage gain is shown in Figure 7.13. A feedback network (470 and 22 kΩ) is used, and the input to this amplifier is the noninverting input of the op amp.

The expression for voltage gain, shown in the figure, involves the ratio of the feedback resistors, but notice that 1 is added to it. For the amplifier of Figure 7.13, 1 is added to 470kΩ/22 kΩ-21 to predict a voltage gain of 22.

As a simple demonstration, use a 0-V input (simply ground the noninverting input). The offset voltage of −13 mV will be used as the input voltage and this will make the output voltage of the op amp swing

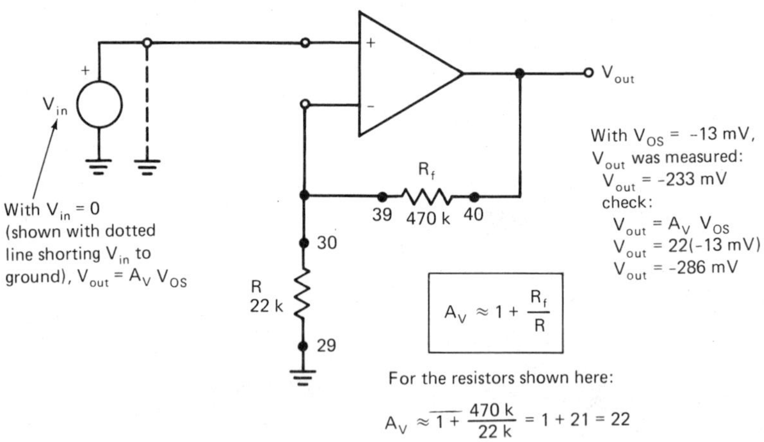

Figure 7.13 The noninverting gain application.

slightly negative, because this is a noninverting amplifier. With this amount of offset voltage, an output voltage of −233 mV was measured. As a check, multiply the −13 mV offset voltage by the predicted voltage gain of 22 to provide −286 mV, a value close to the actual measured output voltage. Errors exist here because the actual measured values of the feedback resistors are not being used and there is also a gain error.

An amplifier with inverting gain

The application circuit of Figure 7.14 shows the very commonly used inverting gain connection. This provides voltage gain, and the input signal polarity is inverted. A positive input voltage will now provide an amplified negative output voltage and vice versa. The same two feedback resistors are used, but the signal to be amplified is now connected to the left end of the input resistor R_{in}.

In this case, the voltage gain is negative and is approximately given by the value of the feedback resistor R_F divided by the value of the input resistor R_{in}. This provides $-(470\text{k}\Omega/22\ \text{k}\Omega)$, or −21.

The large value of the open-loop voltage gain of an op amp implies that the inverting and noninverting input voltages, directly at the op amp, are never very far apart. This gives rise to the very useful concept of a "virtual ground" that exists at the inverting input of an op amp, as shown on the figure. The voltage at the inverting input will not be far from ground because only a very small voltage is needed at this input to supply any required output voltage from the op amp.

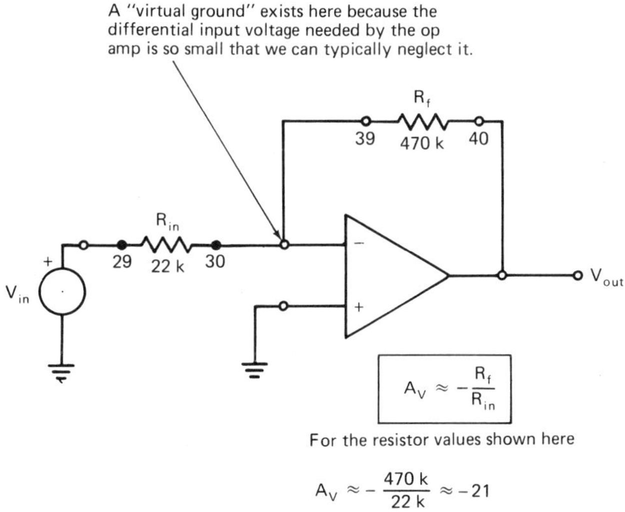

Figure 7.14 The inverting gain application.

Therefore, the inverting input will be "virtually" at the same voltage as the noninverting input: ground. This concept of virtual ground is very useful in op amp application circuits, such as the multi-input summing amplifier that will now be considered.

A multi-input summing amplifier

Virtual ground is the key to the multi-input summing amplifier that is shown in Figure 7.15. The two (or more) input voltages V_1 and V_2 will not interact or interfere with each other because the virtual ground keeps 0 V at the common junction of the input resistors. The first input signal V_1 supplies one component of the input current I_1, and the second input signal V_2 supplies a second component of input current I_2. Both of these input currents can be of either sign, corresponding to either polarity of the input voltages.

The algebraic sum of these input currents flows through R_3 and into the output of the op amp. This creates an output voltage because of the resulting voltage drop across the feedback resistor R_3. The equation for this output voltage is shown in the figure.

A voltage sweep circuit

Figure 7.16 shows an op amp configured as a voltage sweep circuit. This is a popular application for op amps. A 100-μF capacitor is used from the output of the op amp to the inverting input. The Code key

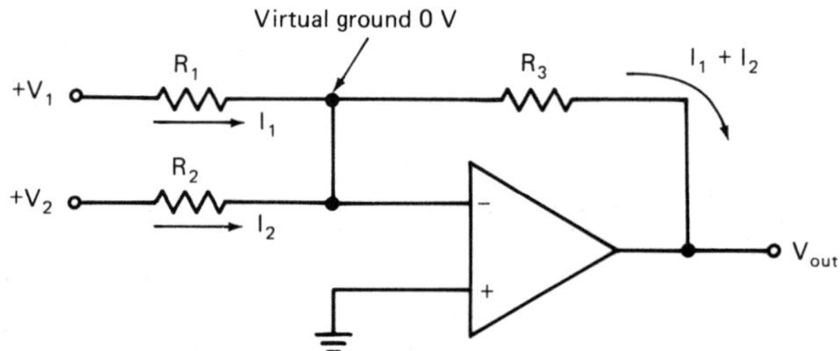

$$V_{out} = -(I_1 + I_2)\,R_3$$

$$= -\left(\frac{V_1}{R_1} + \frac{V_2}{R_2}\right) R_3$$

or

$$V_{out} = A_1(V_1) + A_2(V_2)$$

where $A_1 = -\dfrac{R_3}{R_1}$ and $A_2 = -\dfrac{R_3}{R_2}$

Figure 7.15 The multi-input summing amplifier.

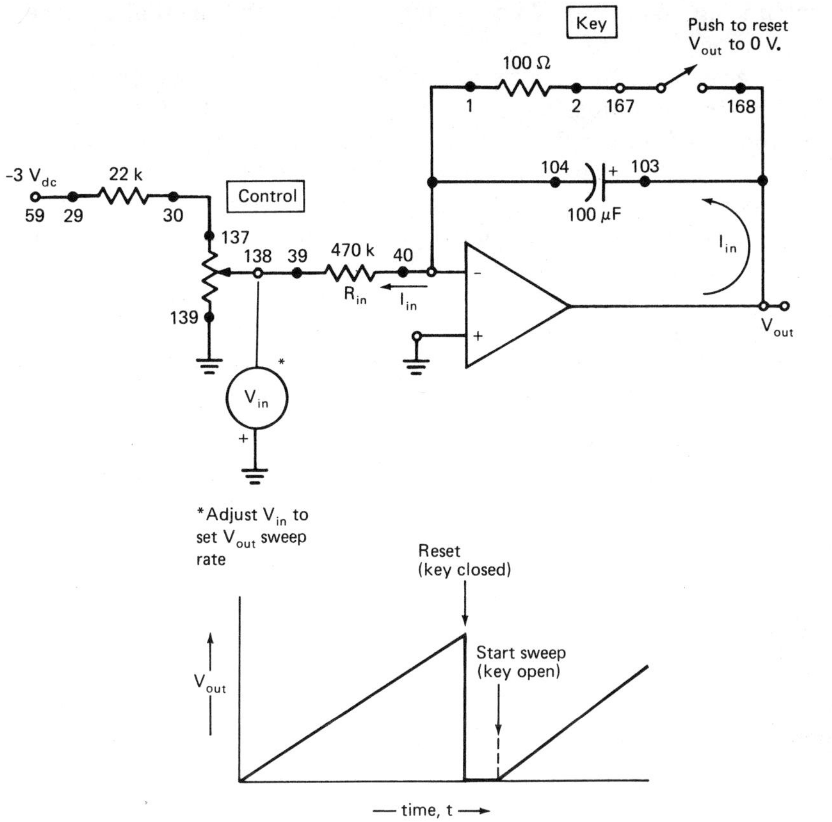

Figure 7.16 A voltage sweep circuit, the integrator.

switch (in series with a 100-Ω resistor) is placed across this capacitor to allow removing the charge from this capacitor and thereby resetting the sweep voltage to zero (whenever the Code key switch is closed).

An input voltage is supplied from the Control circuit that is shown on the left side of this figure. The magnitude of this dc input voltage will control the sweep rate (how rapidly the output voltage changes). This is an inverting amplifier, so a negative dc input voltage is used to provide a positive-going output sweep voltage.

Apply a small input voltage V_{in} of -1 V or -0.5 V. Close the Code key switch to provide an initial sweep output voltage that is very close to zero. When the Code key switch is opened, the output voltage will start a positive sweep. The op amp will saturate when the output voltage gets near 4 V and the output voltage will then stay at this dc voltage level. Any time the Code key switch is closed, the sweep voltage will be reset to zero, and the sweep will restart whenever the Code key switch is again opened. Vary the input voltage and see that the sweep rate changes.

This circuit is also making use of the concept of virtual ground. The negative input voltage is establishing a current flow I_{in} (with the direction indicated on the figure) through the 470-kΩ input resistor. This current is supplied by the output of the op amp and flows through (charges) the capacitor. Even though the voltage across the capacitor changes, the charging current to the capacitor remains constant. This is the basic mechanism that provides a linear output voltage sweep.

This sweep circuit is called an integrator because the capacitor accumulates or integrates the input voltage waveform. Providing a linear voltage sweep is only one use for an integrator circuit.

A square-wave oscillator

A square-wave oscillator can be hooked up, as shown in Figure 7.17. This application circuit only requires three resistors and one capacitor. It's important to connect the 100-μF capacitor to the −4.5-V power supply to prevent a voltage polarity change across this electrolytic capacitor.

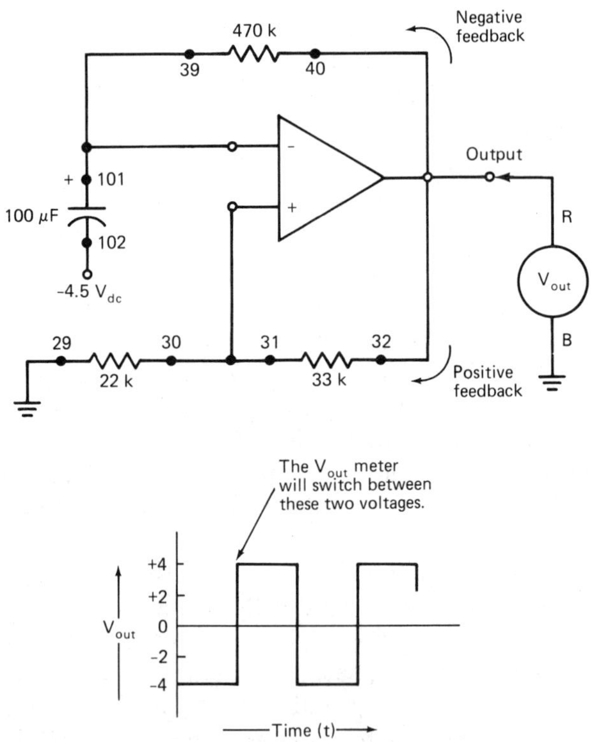

Figure 7.17 A square-wave oscillator.

This is an interesting circuit because two feedback resistors are tied to the output. Both positive and negative feedback resistors are used. The 470-kΩ resistor is feeding back to the inverting input and is therefore providing negative feedback. The 33-kΩ resistor is tied back to the positive input and is therefore providing positive feedback. It's this positive feedback that squares up the output voltage waveform.

Connect the dc voltmeter to the output of the op amp and notice that the output voltage will alternate between a positive voltage and a negative voltage. This is the low-frequency square-wave output voltage that this oscillator provides.

The Op Amp within a Series Voltage Regulator

We have seen how an op amp can be used in many application circuits. Op amps are also basic to a number of other linear ICs. For example, an internal op amp is used in every power supply voltage regulator circuit.

The basic diagram of a series voltage regulator is shown in Figure 7.18. The voltage that is fed to the noninverting input is provided by a 7-V reference diode that is biased with a 500-μA current source. The resistors R_1 and R_2 provide a feedback network to amplify this 7-V reference to provide 15 V at the output. Any desired output voltage

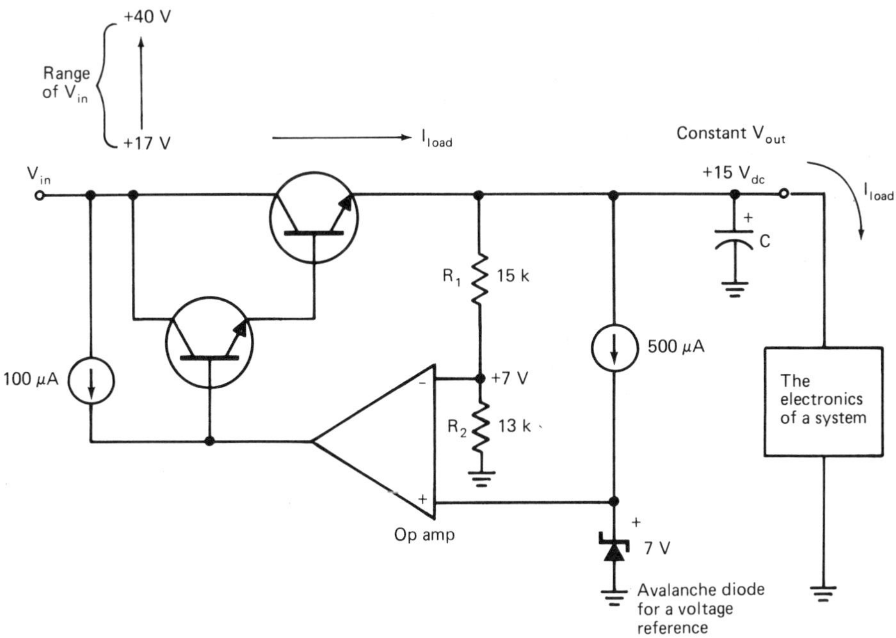

Figure 7.18 How a series voltage regulator makes use of an op amp.

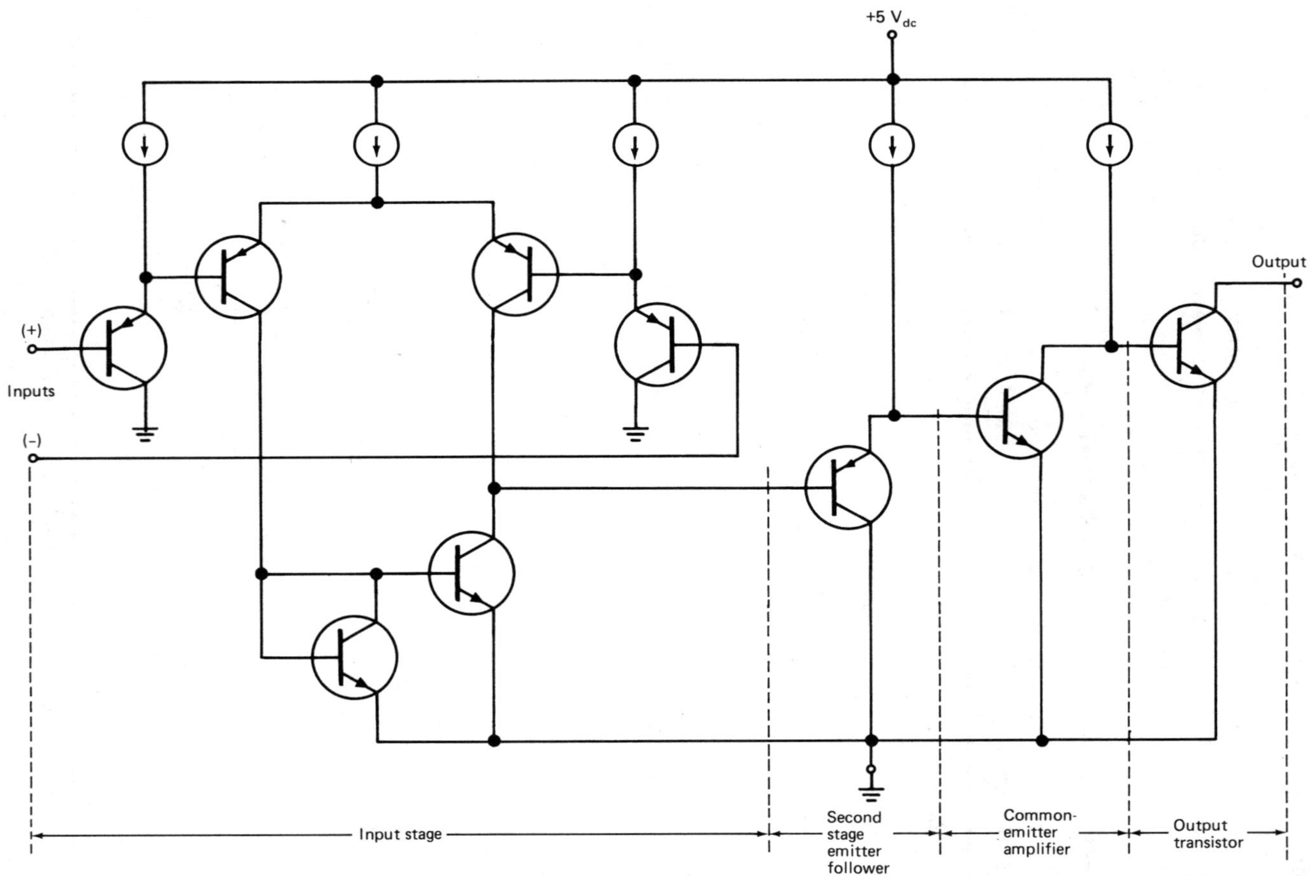

Figure 7.19 A voltage comparator IC (¼ LM339).

can be obtained by appropriately changing these resistors, because they control the closed-loop gain that is amplifying the 7-V dc reference input voltage. The feedback control action of the op amp will always hold the inverting input voltage close to the reference voltage value of 7 V. Because of this, the output voltage stays close to 15 V.

Even if the unregulated input voltage changes (which it will), the output voltage will not be significantly disturbed. This input voltage V_{in} can typically range from 17 (it has to always be a couple of volts larger than the output voltage) to 40 V. V_{in} can also wiggle around (that is, it can contain an ac component of voltage) and a constant 15-V dc output will still be provided. This is called line regulation in a voltage regulator. Ideally, the output voltage should not respond to changes in the input voltage (also called the "line" voltage).

The output of the op amp in this circuit drives a Darlington-connected pair of power transistors so that a large output or load current can be obtained from the voltage regulator. Load regulation is a way to measure how well the voltage regulator is able to supply any load current that may be needed, without allowing the output voltage to change. Good line and load regulation is achieved by making use of a high-gain internal op amp.

The Similarity of an Op Amp and a Voltage Comparator Circuit

You can now understand the basic circuit diagram of a popular voltage comparator: one section of an LM339 quad comparator, as shown in Figure 7.19. This circuit is very similar to an op amp. It has differential inputs and uses a Darlington PNP differential input stage. A current mirror provides the load for this stage. A second stage emitter follower then drives a common-emitter amplifier, and the output of this common-emitter amplifier drives the base of the common-emitter output transistor.

The output stage of this voltage comparator is very simple. It is just a single transistor and the collector of this transistor is the output. This output transistor is either on or off depending on the voltages that are applied at the inputs. For example, if V_{in+} is greater than V_{in-}, the output transistor will be off, and vice versa. Voltage comparators are used in digital types of applications and are usually faster in response time because an on-chip frequency compensation capacitor is not used.

With this background in linear circuits and systems, we will now take a look at what can be expected for the future of analog electronics.

Chapter

8

The Future of Analog Electronics

Electronics started with analog circuits. The telephone and AM radio are two very popular examples. In addition, the first computers were also *analog* computers.

There is strong competition within electronics because an electronic system can be designed using any one, or a combination, of three basic approaches: analog, using op amps; digital, using ON/OFF circuits; or sampled data, which lies between the first two.

Analog circuits have input and output voltages that are continuously available. In contrast with this, digital circuits have inputs and outputs that are digital "words" (a series of 1s and 0s) and these signals are only available at discrete times. A sampled-data circuit takes periodic samples, or "snapshots," of the analog signals and provides the appropriate output signals from these input voltage samples.

In recent years, there has been a continuing encroachment by digital and sampled-data systems into many traditionally analog systems. There have been a number of reasons for this. Digital ICs are steadily decreasing in cost and increasing in performance. This has caused a trend to increase the performance and therefore the complexity of all electronic systems. Many previously analog systems, which usually employ less circuitry but usually require critical-valued components, adjustments, and tougher field service requirements, are moving to a digital solution. The relative ease with which the design and the IC layouts of digital circuits can be automated is also favoring digital systems for the very large-scale integrated circuit (VLSI) era.

This trend to digitize everything is expected to culminate in digital signal processing (DSP) systems. DSP is based on digitizing all analog input signals as quickly as possible and then accomplishing all signal processing such as filtering, finding rates of change, and all types of complex signal processing by performing computations on the digitized samples of the input voltages.

The general pervasiveness of electronics is creating a steadily increasing demand for all semiconductor products. Therefore, any analog losses that result from shifts to a digital system approach are masked by this continually increasing overall demand. For example, every digital system has at least one, and often a few, analog voltage regulators to power the ICs, even though there may be only digital circuitry within the rest of the system. Other linear circuits are also associated with special input and output circuitry that is often associated with a digital system (especially a digital control system). Many linear circuits are required to allow people to talk to and listen to computers. Analog circuits are also always required in the general area of analog signal conditioning to provide the operations that are required to be performed on a low-level analog signal voltage prior to converting this voltage into a digital format (a digital "word" or a number).

The challenge for the analog IC designers is to develop a collection of "library," or building-block, analog circuits that can be added to a complex custom or semicustom CMOS digital IC chip to support the VLSI systems of the future.

Glossary: Electronic Component Symbols

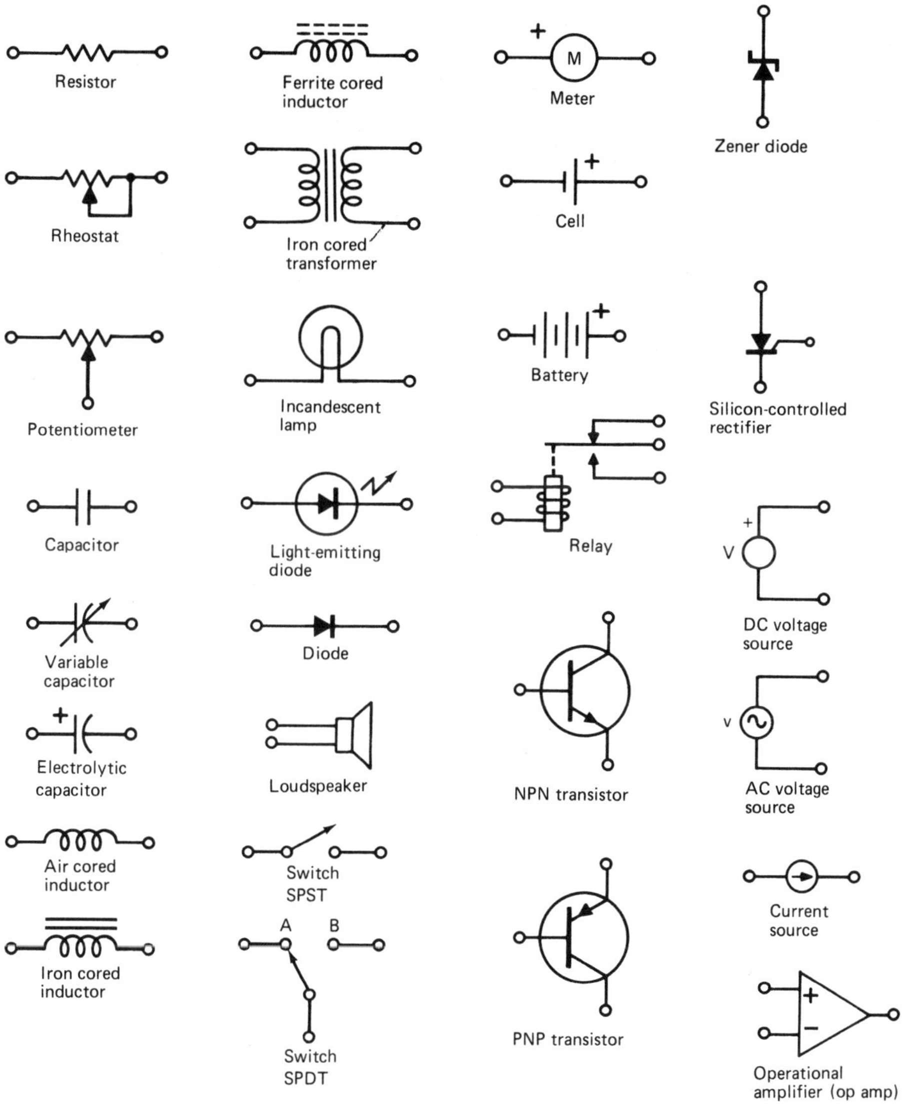

Glossary: Metric Prefixes, Math Symbols, and Powers of 10

Metric prefixes:

For large numbers
- Kilo, k, 10^3
- Mega, M, 10^6

For small numbers
- milli, m, 10^{-3}
- micro, μ, 10^{-6}
- nano, n, 10^{-9}
- pico, p, 10^{-12}

Powers of 10:

$10^3 = 1000.000$
$10^2 = 100.000$
$10^1 = 10.000$
$10^0 = 1.000$
$10^{-1} = 0.100$
$10^{-2} = 0.010$
$10^{-3} = 0.001$

Mathematical symbols

$=$ is equal to
$\approx$ is approximately equal to
A^2 A squared = A X A
$\sqrt{A}$ square-root of A (as $\sqrt{100} = 10$)
$\pi = 3.14$

Glossary: Electronic Formulas

Ohm's law for dc circuits

V = voltage, in volts, V
I = current, in amperes, A
R = resistance, in ohms, Ω
$V = IR$
$I = V/R$
$R = V/I$

Power

P = power, in watts, W
$P = VI$
$P = V^2/R$
$P = I^2R$

Reactance (ac circuits)

X_C = capacitive reactance, in ohms, Ω

$$X_C = \frac{1}{2\pi fC}$$

X_L = inductive reactance, in ohms, Ω
$X_L = 2\pi fL$

Combining components

Resistors R_1 and R_2

In parallel $$R_T = \frac{R_1 R_2}{R_1 + R_2}$$

In series $$R_T = R_1 + R_2$$

Capacitors C_1 and C_2

In parallel $$C_T = C_1 + C_2$$

In series $$C_T = \frac{C_1 C_2}{C_1 + C_2}$$

Inductors L_1 and L_2

In parallel $$L_T = \frac{L_1 L_2}{L_1 + L_2}$$

In series $$L_T = L_1 + L_2$$

Voltage dividers

Resistive

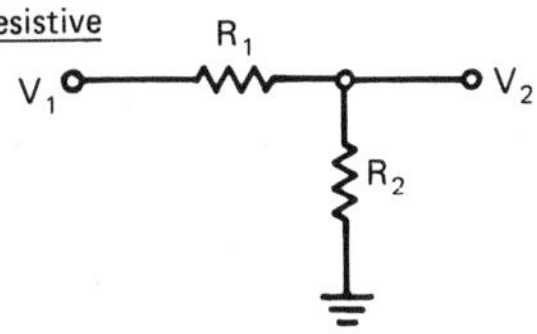

$$V_2 = V_1 \left(\frac{R_2}{R_1 + R_2} \right)$$

Capacitive

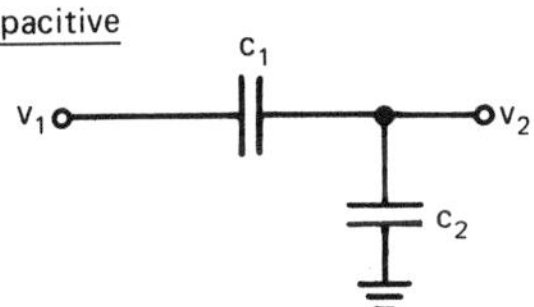

$$V_2 = V_1 \left(\frac{C_1}{C_1 + C_2} \right)$$

Glossary: Measurements Using the Digital Multimeter

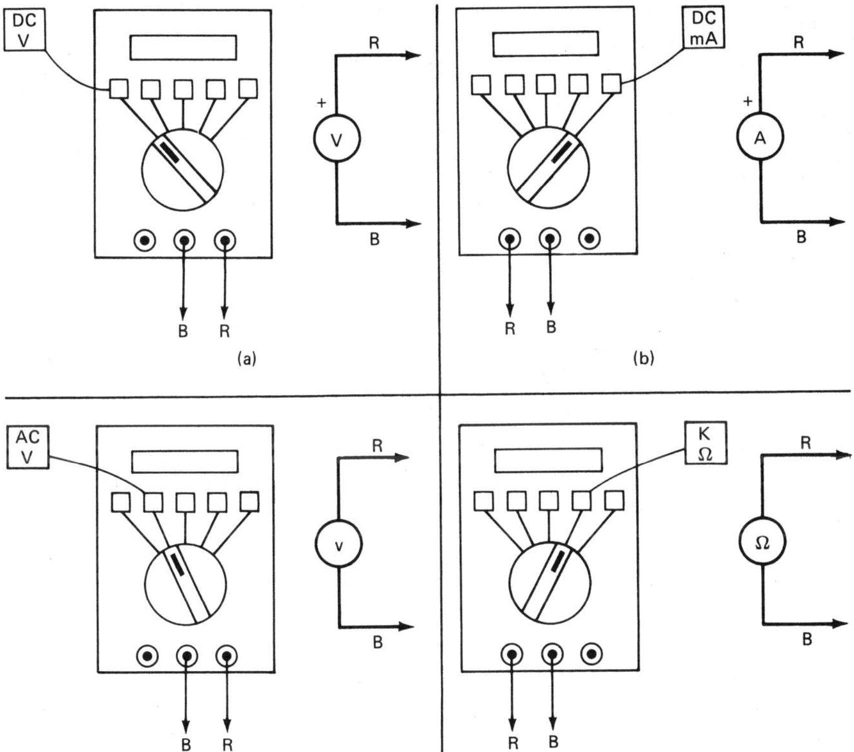

Glossary: Definitions of the Technical Words

active devices Transistors and diodes.

alternating current (ac) A designation of voltages and currents that alternately change sign (if voltage) or direction of flow (if current).

ampere (A) The basic unit of current flow; 1 A is the movement of 1 C of charge past a fixed point each second.

analog A synonym for "linear" in which voltage or current magnitudes represent or are analogous to the magnitudes of measurable things in the physical world, such as pressure, temperature, displacement, humidity, etc.

attenuator A circuit or device that reduces the magnitude of a voltage or a current.

avalanche diode A special semiconductor diode that has been designed to be operated in the voltage breakdown mode to provide a reference voltage. The avalanche mechanism takes place if the breakdown voltage is larger than approximately 5 V; less than this implies a zener breakdown mechanism is occurring within the diode. In the vicinity of 5 V, both breakdown mechanisms can simultaneously take place.

bandpass filter A circuit or device that will allow input voltages over a selected band of frequencies to pass through to the output but will attenuate both lower and higher frequencies that are outside this band.

base One of the three terminals of a bipolar transistor. The base is most often used as the control or input terminal of a transistor.

biasing The act of providing a circuit that will supply suitable dc voltages and currents to a transistor so that this transistor will provide gain or provide a useful circuit function.

bipolar transistor The semiconductor amplifying device that makes use of an alternation of dopant types, such as NPN or PNP, in which the emitter of the forward biased emitter-base junction injects current that is almost all picked up by the reverse-biased collector-base junction.

capacitive reactance (X_c) The numerical measure of the "resistance" a capacitor has to the flow of current when this capacitor is placed across a source of alternating voltage:

$$X_c = \frac{1}{2\pi fC}$$

capacitor A passive electronic component that stores charge and can also be used as a coupling element to block dc voltages and pass ac voltages.

circuit component The very basic building blocks that are used to make up an electronic circuit, such as resistors, capacitors, inductors, transformers, diodes, transistors, switches, etc.

collector One of the terminals of a bipolar transistor that is usually, but not always, associated with the output signal.

common-mode input voltage The component of voltage that is common to both inputs of an op amp (in contrast to the differential-mode input signal).

common-mode rejection ratio (CMRR) The ratio, usually expressed in decibels (db), of the voltage gain of an op amp that is provided for a differential input voltage to the voltage gain that is provided for a common-mode input voltage. Large values of CMRR are desirable. CMRR (numeric ratio) = IO exp [CMRR (db)/20].

conductor A material, usually a metal, that has low resistance to the flow of current.

coulomb (C) The basic unit of electric charge. The smallest possible quantity of charge is that carried by a single electron, -1.602×10^{-19} C.

covalent bond The interatomic bonding that exists in crystals and chemical compounds in which each of two neighboring atoms supply an electron and also share another electron that is supplied by the other atom.

current The flow of charge; 1 A of current flow is the flow of 1 C of charge past a given point each second.

current gain The amount that current is amplified. When in reference to a bipolar transistor, the gain that the transistor has in supplying an amplified collector current in response to a base current.

current source An electronic circuit that will supply a constant value of output current to a load even though the voltage across both the load and the current source changes.

dielectric The insulating material that is placed between the opposing plates of a capacitor.

dielectric constant A numerical measure of the increase in the capacitance value that occurs when a dielectric material (with an associated dielectric constant) replaces air as the dielectric of a capacitor.

digital An electronic system or circuit that is based on the two-state, or binary, number system.

digital signal processing (DSP) An electronic system approach in which all analog signals are digitized as soon as possible, and then all of the required processing of these signals is accomplished by the use of digital computer techniques.

diode A two-terminal electronic component that passes current in one direction but blocks current flow in the opposite direction. A semiconductor diode consists of a PN junction.

direct current (dc) Used to designate either current or voltage that is always in the same direction (current) or of the same polarity (voltage).

emitter One of the terminals of a bipolar transistor. The emitter emits current into the base region and most of this current is then captured by the collector.

farad (F) The unit of capacitance, named in honor of Michael Faraday, that permits the storing of 1 C of charge for each volt of applied voltage.

field-effect transistor (FET) A semiconductor transistor that uses an electric field that is created by a voltage applied to the input terminal (the gate) to affect the conductivity of a channel region that exists between the source and drain terminals. It can be either a junction FET, a JFET, or a metal-oxide-semiconductor (MOS) FET, a MOSFET.

ground The common terminal and often the terminal to which all of the voltages are referenced in an electronic circuit.

hole In doped silicon, the absence of an electron that is created by substituting an atom with three valence electrons, such as boron, into the crystal lattice. This hole acts like a positive charge and therefore is indicated as P-type of doping.

hum In electronics, the interference noise that is caused by undesired pickup of the 60-Hz power-line frequency. This hum can often be heard as the background noise that comes out of a loudspeaker during the quiet times when there is no audio signal.

impedance (Z) The numerical measure of the "resistance" to current flow that a general *RLC* network has when excited with an ac voltage.

inductive reactance (X_L) The numerical measure of the "resistance" an inductor has to the flow of current when this inductor is placed across a source of alternating voltage: $X_L = 2\pi fL$.

inductor An electronic circuit component that stores energy in a magnetic field. Inductors resist changes in the current that flows through them and therefore attempt to maintain this current flow constant

insulator A material that does not allow current to flow when a voltage is placed across it; the opposite of a conductor.

joule (J) In science, the unit of electric energy; 1 J is equal to 1 W of power being dissipated for 1 s (1 W·s)

light-emitting diode (LED) A semiconductor PN junction diode that emits light while conducting a forward current. LEDs have largely replaced incandescent lamps in electronic equipment. There are a number of individual LEDs and a seven-segment LED device in the lab kit used in this book.

linear See analog.

liquid-crystal display (LCD) A solid-state display device in which an applied voltage changes the amount of light that is reflected from the surface of a special crystal. These are low-power displays because they make use of ambient light. The digital multimeter used in this book incorporates an LCD device.

low-pass filter A circuit or device that will allow input voltages with frequencies that are below a predetermined cut-off frequency to pass through to the output but will attenuate higher frequencies.

N-type region The result of doping a silicon crystal with atoms that have five valence electrons, such as antimony or phosphorus. The resulting excess free electrons that result carry a negative charge, and this gives rise to the designation N-type.

negative feedback The act of comparing a fraction of the output signal to the input signal at the input to an amplifier in such a way that the amplifier will keep this fraction of the output signal always exactly the same as the input signal. This action increases the gain predictability and reduces distortion, at the expense of providing a smaller overall amplification of the input signal than could be provided if negative feedback were not used.

no connection (NC) Used on an electronic schematic diagram to indicate that there is no electrical connection made to the terminals that are labeled NC.

offset voltage The undesired, small, nonzero amount of input difference voltage that must be supplied directly at the input of an op amp to cause the output voltage to go to 0 V.

ohm (Ω) The basic unit of resistance, named in honor of George S. Ohm; 1 V applied to a resistance of 1 Ω will cause 1 A of current to flow.

Ohm's law The relationship between voltage V, resistance R, and current I in a dc circuit that is expressed as $V = IR$.

operational amplifier (op amp) The basic gain block of linear systems. This versatile amplifier can be configured to perform basic mathematical operations (and thus the name op amp was coined) on input signals and, in addition, can also be used in many ingenious application circuits.

oscillator An electronic circuit that supplies an output signal without requiring an input signal. Oscillators are used to provide many different types of output voltage waveforms, such as sine waves, square waves, ramps, triangle waves, etc.

oscilloscope A piece of electronic test equipment that is used to display a voltage or current waveform as a function of time.

P-type region The result of doping a silicon crystal with atoms that have three valence electrons, such as boron. The resulting lack of an electron, a hole, behaves like a positive charge and gives rise to the designation P-type.

passive components Electronic circuit components that do not require special biasing to make them operational such as R's, L's, C's as opposed to transistors and diodes.

potentiometer A resistor that includes a sliding, contacting wiper that can sample the voltage continuously from one end of the resistor to the other.

Potentiometers have three terminals and are used as volume controls in radios and TV sets.

quiescent Literally means quiet; in electronics, quiescent refers to the voltages and currents that exist in the quiet state, without a signal applied.

rectifier A circuit used for a diode in which only the positive or negative voltage swings of an ac waveform are allowed to pass through to the rest of the circuitry.

resistor A passive circuit component that is designed to have a specific value of resistance. Resistors are supplied in a wide range of power dissipation capabilities.

resonant circuit An electronic circuit usually consisting of an inductor and a capacitor that is designed to be resonant at a prespecified frequency. They circuits can be used to select a particular signal frequency out of a composite signal spectrum that simultaneously contains many signals at various frequencies.

rheostat A two-terminal, mechanically variable resistor.

root mean square (RMS) Actually, the square root of the mean squared value. The root mean square value of an ac voltage is also a measure of the heating effect that would be obtained if this voltage were connected to a resistor. An ac signal of 10 v_{rms} has the same heating effect as a dc voltage of 10 v_{dc}.

semiconductor A solid material that is neither a good conductor nor a good insulator. Silicon (which has largely replaced germanium) is today's most popular semiconductor material, and there is much interest in gallium arsenide for the future.

shunt A synonym for being connected in parallel.

silicon-controlled rectifier (SCR) A three-terminal power-control semiconductor device that consists of four alternating doped regions, as NPNP or PNPN. Once triggered into conduction, an SCR will permanently latch up, and the voltage must be interrupted to reset this conducting latched state and turn the SCR off. SCRs can exist as undesired parasitic devices in bipolar and CMOS ICs.

sine wave A particular ac voltage or current waveform that follows a mathematical sine wave; a smoothly changing waveform that alternatively reaches a positive peak value and then a negative peak value.

tank circuit A synonym for a resonant circuit; see resonant circuit.

time constant In a simple *RC* circuit, the time constant is the product of the values of the *R* and *C* and has the dimensions of time in seconds. A voltage source, charging a capacitor through a resistor, will supply 63 percent of this voltage to the capacitor in a time interval of one time constant. After three time constants have elapsed, the capacitor will be charged to 95 percent of the voltage of the voltage source.

transducer A device that converts a physical parameter into a voltage or a current or vice versa. A loudspeaker and a microphone are common examples of transducers.

transformer A magnetically coupled set of at least two windings in which the driven winding (the primary) causes a voltage to appear at the output winding (the secondary). Transformers are typically used to increase or decrease the magnitude of an ac voltage. Transformers can also be used for electrical isolation and to alter the impedance of a "transformer-coupled" load.

transistor The popular semiconductor three-terminal amplifying device; see bipolar transistor and also field-effect transistor.

transistor action The mechanism that takes place between two closely spaced diodes in which the forward current of one of these diodes is almost completely stolen by the second, reverse-biased diode.

triac A special power-control semiconductor, similar to an SCR, that is designed to operate with ac excitation. Triacs are used in incandescent light dimmers and motor speed controls.

very large-scale integration (VLSI) The name used for the present complexity of the circuitry that can be placed on a single silicon IC chip.

virtual ground A useful applications concept in the use of op amps in which the inverting input voltage of the op amp can be counted on to remain at "virtually" the same voltage (0 V, or ground) as the ground-connected noninverting input. As a result of the large open-loop voltage gain of an op amp, any desired output voltage can be obtained by such a very small voltage directly at the inverting input that this inverting input operates at voltages that are "virtually" the same as ground.

voltage The pressure in an electronic circuit, measured in volts, that causes current to flow.

voltage comparator An electronic circuit, somewhat similar to an op amp, that is designed to have the output either on or off (that is, the output voltage either low or high) in response to the differential input voltage to the comparator. Comparators are used in digital circuit applications.

voltage source A source of voltage; an ideal voltage source has zero source impedance and can also supply whatever current is demanded by the load that is connected to it. The battery in a car is almost an ideal voltage source (especially if the load current is kept small, say less than about 50 A).

watt (W) The unit of electric power; in a dc circuit, 1 W of power is dissipated in a 1-Ω resistor if 1 V is applied across this resistor.

zener diode See avalanche diode.

Index

ABOUT THE AUTHOR

Thomas M. Frederiksen, engineer, author, and seminar leader, founded his own company, Intuitive IC Seminars, to provide instruction to electronic design engineers and the many nonelectronically educated people who either work for, or with, a high-technology electronic company.

Upon earning his BSEE degree from California Polytechnic State University at San Luis Obispo, Frederiksen started his professional career as a development engineer with the Motorola Systems Development Laboratory. Subsequently he worked with the Microelectronics Group at Hughes Semiconductor Division and later became a senior project engineer at Motorola Semiconductor Products Division. Mr. Frederiksen then joined National Semiconductor Corporation, Santa Clara, California, where he developed custom ICs and standard single-supply building-block circuits for automotive and industrial applications. He designed the popular Quads: LM3900, LM324, and LM339. The LM324 is today's most popular operational amplifier. He has also been involved with analog-to-digital converters, which will interface to microprocessors, and other data acquisition circuits.

Mr. Frederiksen is the author of five books: *Intuitive IC Electronics, Intuitive CMOS Electronics,* and *Intuitive Op Amps* for engineers and technicians and *Intuitive Digital Computer Basics* and *Intuitive Analog Electronics,* which introduce the basics of electronics and also digital and analog circuits and systems, for nonelectronic professionals.

Mr. Frederiksen holds more than 40 patents on linear ICs and devices, has been a frequent contributor to the professional literature, and has given many major seminars on linear ICs for both Motorola and National Semiconductor within the United States and Canada and also abroad. In 1977 he received the International Solid State Circuits Conference Best Paper Award.